BIOTECH AT GRASSROOTS

Biotechnology Research Series

1. Educational Infrastructure for Biotechnology in India
   —*R.K. Mishra, B. Navin, P. Geeta and Ch. Lakshmi Kumari*

2. Venture Capital Financing for Biotechnology
   —*R.K. Mishra*

About the Series

Biotechnology is the term for techniques of managing biological systems for human benefits. The production of varied lifestyles and health products are all the result of brilliant work of biotechnologists. This series attempts at unraveling the penetration and impact of biotechnology in various areas of human life at grassroots level. Its two volumes are already brought out, namely Educational Infrastructure for Biotechnology in India and Venture Capital Financing for Biotechnology. This volume is the third in the series. Another volume on 'Public-Private Partnership in Biotechnology Research and Commercialization' is in the pipeline.

About the Authors

Alok Ranjan Jha is an avid researcher in the area of science and technology for rural upliftment. Presently, he is working for a research and analytics company headquartered at Texas, USA. He has been passionately involved in rural development initiatives and allied research across the country. He is on the board of several grassroots NGOs working in the area of rural/social development and livelihoods.

R.K. Mishra is currently ***Senior Professor*** and ***Director***, Institute of Public Enterprise, Hyderabad. Earlier, he has taught at the University of Bradford, United Kingdom and was a visiting Professor of Maison Des Sciences De L' Hommes, Paris. He has several international and national pubiications to his credit including two on biotech—Educational Infrastructure for Biotechnology in India, and Venture Capital Financing for Biotechnology. He is Vice-President, International Association of Schools and Institutes in Administration (IASIA), Brussels. He is the Chairman of the Working Group II: Public-Private-Partnership and Enterprise Management constituted by IASIA. He is on the Steering Committee of Network of Asia-Pacific Schools and Institutes of Public Administration and Governance (NAPSIPAG) supported by the Asian Development Bank. He is Honorary Member on the Board of Research, American Biographical Institute, USA.

Biotechnology Research Series No. 3

BIOTECH AT GRASSROOTS
Impact Assessment of AgNet Project

Alok Ranjan Jha
R.K. Mishra

CONCEPT PUBLISHING COMPANY, NEW DELHI-110059

ISBN 13: 978-81-8069-629-9 **ISBN 10: 81-8069-629-4**

First Published 2009

Published and Printed by

Concept
Since 1974

Concept Publishing Company

Regd. Office:

A/15-16, Commercial Block, Mohan Garden

New Delhi-110059 (India)

Phones : 25351460, 25351794, *Fax* : 091-11-25357109

Email : publishing@conceptpub.com,
website: www.conceptpub.com

Editorial Office:

H-13, Bali Nagar, New Delhi-110 015, India.

Cataloging in Publication Data--*Courtesy:* D.K. Agencies (P) Ltd. <docinfo@dkagencies.com>

Jha, Alok Ranjan.

Biotech at grassroots : impact assessment of AgNet Project/ Alok Ranjan Jha, R.K. Mishra.

p. cm. -- (Biotechnology research series ; 3)

Study conducted at Nalgonda and Mahaboobnagar districts of Andhra Pradesh, India.

Includes bibliographical references (p.).

Includes index.

ISBN 13: 9788180696299 ISBN 10: 8180696294

1. AgNet Project (India). 2. Agroforestry--Technology transfer--India--Nalgonda (District) 3. Agroforestry--Technology transfer--India--Mahbu?bnagar (District) 4. Forestry biotechnology--India--Nalgonda (District) 5. Forestry biotechnology--India--Mahbu?bnagar (District) 6. Arid regions forestry--India--Nalgonda (District) 7. Arid regions forestry--India--Mahbu?bnagar (District) 8. Sustainable forestry--India--Nalgonda (District) 9. Sustainable forestry--India--Mahbu?bnagar (District) I. Mishra, R. K. (Ram Kumar), 1948-, joint author. II. Title. III. Series: Biotechnology research series ; 3.

DDC 634.99095484 22

To

Our Inspiration

Professor T. Navaneeth Rao

Emeritus President, IPE,
Former Vice Chancellor,
Osmania University

Preface

The innovations on technological front are creating a catalyst for spawning growth in newer dimensions. Despite these technological advances, approximately a third of humanity are deprived of basic needs and lack of access to resources. Biotechnology tools, combined with conventional techniques, will help in the emergence of appropriate technologies, which can contribute to meet the challenges thrown by unequal distribution of resources. For achieving this, biotechnology will have to adopt a people-centric approach in place of a market-centric approach. It can benefit the poorest of the poor when suitable innovations are spawned and developed keeping in focus the needs and constraints of the main stakeholders, the resource poor farmers and the conditions prevailing at the grassroots level.

The present book is the third in the series of books on Biotechnology co-authored by Prof. R.K. Mishra. While the first two books in the series attempted to study the status of Educational Infrastructure for Biotechnology in India and Venture Capital Financing for Biotechnology in Indian context, this volume in the series systematically analyses the impact of biotechnology at grassroots level. The book examines this issue and tries to explore whether the application of biotechnology would help mitigate the sufferings of small and marginal land holders of dryland agriculture. The book aims to explore the role of high technology, specifically, biotechnology, in improving the socio-economic conditions of rural poor through sustainable livelihoods approach. For this the book documents the ideation and execution mechanism of AgNet and tries to assess its social impact on the livelihoods of rural peasants. Furthermore, book explores the possibility of upscaling AgNet and similar technology transfers and suggests policy measures for sustainable livelihoods through biotechnology.

Conceptually, for the research, book draws from both the traditions of research approach, i.e., positivist and social constructivist.

The research has taken some elements from positivist tradition as it involves natural sciences and is concerned with finding out the objective truths. At the same time, it has also brought in the elements of social constructivist tradition, as it is people-centred.

Based on the nature of research questions and the specific objectives, the study followed descriptive research design. The main aim of this design is to present the situation and status of technology adoption at grassroots level. The general methodology of the proposed study included an intensive, participatory and on-the-field-study and series of consultations at local, and district levels.

The study findings are that on the whole, AgNet project has been able to achieve a social impact. Specifically, it was very successful in generating and spreading awareness about agroecological approach for the benefit of small and marginal farmers of the dryland tracts. It also has achieved success in the diffusion of technology by removing and minimizing constraints for the adoption of specific agro-forestry practices. But the project could not achieve similar success in forging relationships and increasing the social capital thereby. Also, it failed to decrease discrimination by caste, gender and economic class, increase the participation in village development activities and increase awareness about their rights and responsibilities. In the light of the findings it can be conclusively said that this project can be replicated to other dryland tracts. In fact it has become a successful template for the replication for sound agro-ecological practices.

Analysis on Sustainable Livelihoods Framework suggests that overall the project has become successful in decreasing the vulnerability context of the participants by the innovative use of recent advances in agricultural technology through increasing human, social and natural capital which have in turn resulted in the increase of financial capital.

Finally, the book recommends that AgNet project should explore further linkages and possible integration of carbon-trading as a secondary benefit to the participants to ensure sustainability of the effort; a study mainly focusing on developing a framework for measurement of social capital at the project site should be taken up to understand the changes in social capital as effected by the specific project; a baseline study of the participants and the project site should be made at the time of initiation of any project; for the projects that are taken up a log-frame including the criteria and indicators should be in place for effective monitoring, evaluation and impact assessment; a concerted and whole-hearted effort should go in the formation, nurturing and developing of CBOs for ensuring the sustainability of the project; a robust set of criteria should be laid down and strictly adhered to for the selection of beneficiaries, NGOs and the project sites; the project should adopt a pro-poor bias in its approach and make sure that the benefits of the efforts are channelled to poor and the poorest of poor; the project should have a robust training programme adapted to the needs and cultural settings of the intended beneficiaries and the project should put in enough efforts and direct its energies not only towards technology-transfer but also towards the overall empowerment and upliftment of the beneficiaries by adopting a holistic approach.

Alok Ranjan Jha

R.K. Mishra

Acknowledgements

We heartily appreciate the ingenuity of APNLBP for promoting innovative applications of recent technological advancements and thank them for the same.

We are thankful to Shri V. Anji Raju, Shri P. Gopikrishna, Dr. P.S. Janaki Krishna, Dr. M.L.N. Rao and to Dr. G. Vijay for their useful and incisive insights during our interactions.

We are grateful to Shri S. Sudhakara Chary and survey team members for their sincere and unflinching commitment during the research phase.

We also thank the NGO personnel for their excellent co-operation during field study. A thanks is also due to our students especially Ms. D. Suman (PGDBM 2008) for the assistance.

We extend our sincere thankfulness to the generous help and hospitality of all the respondents, stakeholders and other villagers with whom the team interacted and gained a first hand understanding of the dynamics of rural setting.

Finally, we are most appreciative of the support of the management and staff of the Concept Publishing Company, especially Shri Ashok Kumar Mittal for extending their excellent support to us.

Alok Ranjan Jha

R.K. Mishra

Contents

1

Introduction

All across the globe, the relentless innovations in technological front are creating a catalyst for spawning growth in newer dimensions. The recent advances made in high technology sectors have opened the vistas of unimaginable opportunities for the humankind. It can be seen that new frontiers in technology are now driving most economic development across countries. Recent research works buttress the fact that "*Technological knowledge accumulation is now being widely acknowledged and acclaimed as a source of economic growth*" (Singh, 2006). Globally, the trade in high technology has increased by 13.19 per cent from 1985 to 2000 (World Bank, 2004).

Despite these technological advances and ever increasing prosperity, the global inequality has increased. A recent World Trade Organisation publication states that: "It is an empirical fact that the income gap between poor and rich countries has increased in recent decades" (Nordström, 2000, p.3). In the *World Development Report 2000*, it was suggested that "*... income inequality between countries has increased sharply over the past 40 years*" (World Bank, 2000, p.51). The United Nations Development Programme said in its 1999 *Human Development Report*: "*Inequality between countries has ... increased*" (UNDP, 1999, p.3). This is clearly discernible with the fact that in 2001, 21.1 per cent of the world population, i.e., 1092.7 million people, was living below a minimal income level of $1.08 per day; whereas more than half of humanity (52.9 %), i.e., 2735.6 million people were living below an income level of $2.15 per day in 2001 (Chen and Ravallion, 2005).

Taking the aforementioned facts, it can be discerned that a vast majority of global population is deprived of basic needs and lack of access to resources. Along with difficulties in earning a decent livelihood and foodsecurity, these resource poor are the primary victims of the impact of climate change on the environment and incidence of natural disasters. In the light of these discomforting facts, it can be said that although technological development and knowledge economy have added to prosperity but these have been glaringly unable to help address the social and economic upliftment of masses of have-nots and redressing their most pressing issues of concern. So, it seems indirectly, knowledge based high technologies have acted as instruments of enhancing global inequality and added to it another dimension of technological inequality.

Among other causes of growing technical inequality, the most important one remains the lack of initiative and innovative ideas of applying recent advancements in high technology for the poor and the poorest of the poor. Added to that, application of technology remains an ignored and under-researched element for bettering the condition of this vast expanse of humanity. Another important cause is that technological products are primarily researched, developed, and marketed by the private sector. Exclusivity of ownership of technology traits often times encourages private-sector companies to influence the input markets. In order to bridge the widening gap between rich and the poorest of the poor, the starting point must be tackling the most pressing issues faced by the disadvantaged with the help of advancements in science and technological breakthroughs.

By 2030, the world's population is expected to grow to 8.1 billion at a rate of over 75 million people per year. Almost all of the population increases will occur in developing countries (FAO, 2000) that can ill-afford additional population pressures. Based on a population-increase-only projection with per capita consumption remaining constant, world cereal production must rise from approximately 1.92 billion tons in 1990 to 2.88 billion in 2030 to match the demand. Although the number of undernourished people in developing countries is expected to decrease, the global food-system situation will continue to be unacceptable. Some authors (Moore-Lappé *et al.,* 1998) maintain that there is currently enough food to feed the world's population adequately; they argue the problem of world hunger is not one of quantity but of unequal distribution. However, even if we resolve the issue of distribution in the short-run, the future growth in food demand will require increases in productivity from a decreasing stock of arable land. The challenge, therefore, is not only to feed more people, but to do so with less available arable land, fewer non-renewable resources, less water, and fewer people engaged in primary agriculture (Conway and Ruttan, 1999).

These population and resource facts, combined with a renewed commitment to fighting poverty, indicate that the main thrust of national and international policies aimed at solving issues of rural poverty and food insecurity must include broader agricultural and rural development objectives, such as significant increases in local food production. To escape from poverty, rural population in developing countries depend directly or indirectly on increased agricultural productivity. Because the rural poor represent a significant percentage of the total population in developing countries, an innovation that increases productivity will have a major impact on food-security efforts and a nation's poverty.

Agriculture is pivotal to poverty reduction, food security, holistic development and environmental sustainability. It is undergoing a major shift within many developing countries, from mere subsistence approach to brazen corporatisation by multinational companies. At the same time, increasing numbers of people lack access to affordable nutritious food. As FAO points out "an additional 2 billion people will have to be fed over the next 30 years from an increasingly fragile natural resource base. More than 842 million people are chronically hungry, most of them in rural areas of poor countries, and billions suffer from micronutrient deficiencies, an insidious form of malnutrition caused by the poor quality of food and lack of diversity in their habitual diet" (FAO, 2003).

The powerful technology base, such as biotechnology, combined with the development of enhancing technologies, can contribute to meet these challenges thrown by unequal distribution of not only resources but also the technical know-how. The Convention on Biological Diversity (CBD) defines biotechnology as: "any technological application that uses biological systems, living organisms, or derivatives thereof, to make or modify products for specific use" (Secretariat of the Convention on Biological Diversity, 1992). This definition includes medical and industrial applications as well as many of the tools and techniques that are used most often in agriculture and food production.

Biotechnology can overcome production constraints that are more difficult or intractable with conventional breeding. It can speed up conventional breeding programmes and provide farmers with disease-free planting materials. It can create crops that resist pests and diseases, replacing toxic chemicals that harm the environment and human health, and it can provide diagnostic tools and vaccines that help control devastating animal diseases. It can improve the nutritional quality of staple foods such as rice and create new products for health and industrial uses.

It has been said that biotechnology dovetails to the needs of developing countries. Biotechnologies that are focused on small holder problems, undertaken in an integrated manner and along with traditional research aimed at improving agronomic practices, can help poor farmers increase productivity (Pinstrup-Andersen, Pandya-Lorch, and Rosegrant, 1999). Thus, while only part of a total solution that involves better market conditions, policies, and access to production resources, biotechnology can contribute to addressing poverty issues in developing countries.

Also, biotechnology possesses remarkable adaptability to local conditions. But most of the current advances in modern biotechnology research and development are taking place in the developed countries, mainly under the initiative of the private sector catering to the demands of resource reach populace of north. If biotechnology has to benefit the millions of poor and poorest of the poor of south it will have to adopt a people centric approach in place of a market centric approach. Biotechnology can benefit the poor and the poorest of the poor when suitable innovations are developed keeping in focus the needs and constraints of grassroots conditions. Also, it will be helpful when poor farmers in developing and least developed countries have access to them so that it can bring socio-economic enhancement in their status.

A few organisations and private players interested in developing the technology and its application for holistic development of the masses, point to the benefits of biotechnology for farmers, the biosafety approval process, as well as the technology's potential for addressing pressing problems of food insecurity and livelihoods in developing countries. Their claims have been validated by some excellent reviews of agronomic and socio-economic studies, e.g., Marra, Pardey, and Alston (2002); Shelton, Zhao, and Roush (2002); and the Commission of the European Communities (2002). The main conclusion of these studies is that biotechnology varieties provide significant benefits to farmers. For example, farmers can use less pesticide or substitute it with less toxic ingredients. In some cases, the use of biotechnology may facilitate the adoption of erosion conservation methods, such as the no-tillage or reduced-tillage practices. In most cases, the benefits for the farmer outweigh the increased investments they have to make in the form of usage fees for the developers of the technology.

The present study is an attempt to examine whether India, a developing country with a superior high technology base, which boasts of emerging as a leading global player in this technology, can also show the way for improving the lives of millions of rural poor by the innovative application through this miraculous tool of science. The Andhra Pradesh Netherlands Biotechnology Programme is one such increasingly diminishing attempt which strives to make a qualitative change in the lives of rural farmers through the innovative and participatory application of advances in biotechnology.

2

Context of the Study

Literature Review

A comprehensive study of socio-economic impact of bio-technologies would have look into its impact on changes to:

(a) agrarian productivity, and
(b) agrarian relations.

R. Radhakrishna (2002) argues that India's agricultural growth reveals two distinct phases: (i) pre-green revolution period, and (ii) post-green revolution period. While the pre-green revolution period had significant imbalance between food production and demand for food, green revolution from mid-1960s to the close of 1980s achieved near self-sufficiency in production of food and consequently food security. This phase, however, threw up new problems of inter and intra-regional and inter-crop imbalances in crop yields.

Tracing the impact of changes in agrarian economy on the agrarian relations, scholars have argued that as part of the green revolution, when land was becoming an increasingly productive asset, because of both public investments and new technologies, land reforms were subverted with impunity. The ceiling surpluses were kept to the minimum and the surplus land surrendered was often of very poor quality…what is retained by the rich peasantry is not only better quality land but also the land that is a better productive asset because of the benefits of State investments in providing infrastructure like irrigation and power facilities along with the benefits of heavy doses of subsidies (D. Narasimha Reddy, 2002).

Thus, when we refer to the top 10 per cent of the rural households operating about 50 per cent of the land, a substantive proportion of this land is more productive land, while the small, marginal farmers constitute about 80 per cent of the households operating only about 30 per cent of the land, a substantive part of which is likely to be sub-standard (*ibid.*).

While land reforms failed to bring about any radical redistribution of land, the introduction of new technology has brought about a sharper differentiation among the peasantry by opening up more profitable opportunities in agriculture to the rich peasantry. The new technologies have been biased in favour of those who have better command over resources (*ibid.*).

In the post-liberalization phase, these impressions of a bias towards the well-off social groups being benefited continues. Changing consumption patterns driven by rising income levels, changing relative prices between cereals and pulses and high value agriculture, increased urbanization and a more liberalized trade regime are contributing to a shift from cereal led growth to high valued agriculture (Ashok Gulati and Kathleen Mullen, 2003). However, commercialization of agriculture has been biased towards farmers who have access to capital and has led to focus on regions where irrigation facilities are abundantly available.

'The Green Revolution triggered off the first large scale rural migration in India wherein the poorly paid under-employed labour in the rainfed marginal environment started migrating to the green revolution regions.' It is pointed out that 'the rising real wages led to labour substitution in irrigated agriculture and made rainfed farming un-remunerative' (R. Radhakrishna, 2002).

Thus, the resource scarce farming communities and regions have suffered marginalization and exclusion in terms of policy focus. While for resource rich farmers and regions liberalization has been seen as an opportunity, for the resource scarce farmers and regions the challenge is to attain sustainable livelihoods and incomes to prevent these marginalized farming households from reclining to below poverty line levels of household incomes. The low and medium bio-technologies address issues of income augmentation amongst these marginalized groups and regions.

Although, the lower and middle level bio-technologies may not be seen as technologies contributing towards maximization of output, these technologies are rather to be analyzed as methods of generating sustainable livelihoods under conditions of low resource availability. (Jose Falck Zepeda, Joel Cohen, Ruth Meizen Dick, and John Komen, 2002). Scholars have argued that poverty amongst poor farming households are closely interlinked with productivity levels in agriculture. Bio-technologies that are focused on small-holder problems undertaken in an integrated manner, and along with traditional research aimed at improving agronomic practices, can help poor farmers increase productivity (Pinstrup Andersen, Pandya Lorch and Rosegrant, 1999).

In the recent past agriculture has become extremely risky. This is owing to bad and uncertain returns. Agricultural prices both in international and national markets have been identified as exhibiting price volatility. The reasons for price instability has, however, not been very clear. Thus, increased productivity levels may not always be a solution under these circumstances. Rather, cost reduction strategies under these conditions becomes a prerequisite for farmers to cope with the increasing levels of risk and uncertainties in the market (C.S.C. Sekhar, 2003). The low and medium level bio-technologies can be effective cost cutting strategies.

Given this background, bio-technologies have great relevance in addressing these several socio-economic issues. Addressing these issues would suggest impact on distribution of the benefits of development in favour of the sections that have not been beneficiaries of the green revolution.

It is observed that bio-technologies are adoptable by small and marginal farmers also as they are less expensive and less resource intensive in terms of need for supportive resources to make these low and medium level bio-technologies usable (Florence M. Wambugu and Sam Wakhusama, 2002). The introduction of these bio-technologies have resulted in reduced dependence on credit markets, have been gender neutral in terms of having no gender bias in the employment opportunities generated consequentially and have in fact been empowering in terms of greater control of women over production processes. In the same work, reduction in child labour as a consequence to introduction of bio-technologies is identified (*ibid.*). It is pointed out that one of the major obstacles to technological upgradation in agriculture is the lack of markets (p.89). Technological change could be seen under these circumstances as a linear process. Shift to lower and middle levels of bio-technology would be more realistic before any shift to higher levels of bio-technology. Higher levels of bio-technologies are caught in a controversy due to ignorance about the possible future effects on ecology and health (Andre C. Fish, and Larisa Rudenko, 2004: 21-24). Over the years these issues may be addressed with scientific evidence emerging from the ground. The lower and middle levels of bio-technology however, are less controversial and have a notable impact on productivity. These technologies have greater indigenous possibilities of innovation as well as easier accessibility to farmers since these are closely linked with traditional practices (Vandana Shiva, Afsar H. Jafri, Gitanjali Bedi, and Radha Holla-Bhar, 1997: 4). It is, therefore, argued by some scholars that farmers with low levels of literacy can pick up the skills necessary for application of bio-technologies (Florence M. Wambugu and Sam Wakhusama, 2002). It thus helps address constraints to percolation of technologies faced by the rural poor of the Third World Countries.

Relevance of Biotechnology to Site of Research

This study aims at investigating into the socio-economic impact of bio-technologies in two districts, viz., Nalgonda and Mahaboobnagar located in Telangana region of Andhra Pradesh. All the farmers who have adopted these technologies are small and marginal farmers.

Mahaboobnagar district has a total area of 18400 sq. kms with a total population of 3.5 million persons. 90 per cent of the total population in Mahaboobnagar lives in rural areas. 39 per cent of the total main workers were engaged in cultivation in Mahaboobnagar in 1991 this has decreased to 36 per cent by 2001. Nalgonda has a total area of 14200 sq. kms and a total population of 3.2 million persons. 87 per cent of the total population in Nalgonda live in rural areas. 32 per cent of the total main workers were engaged in cultivation in 1991. This decreased marginally to 30 per cent by 2001.

In terms of the operational-holdings, small and marginal farmers constituted about 80 per cent of the total number of farmers. These farmers were holding a total of 34 per cent of the total area cultivated (S. Sudhakar Reddy, S. Galab and P. Padmanabha Rao 2003: pp. 444).

For the State as a whole, the percentage of cultivators has decreased from 12.53 per cent in 1987-88 to 9.67 per cent by 1993-94. It is maintained by the scholars that while shift of agricultural labour away from agriculture to non-farm sector represents a pull factor, such a shift in case of cultivators represents push factor. Thus, clearly a distress in agriculture can be seen in the reduced percentage of cultivators (S. Mahendra Dev and Vijay Mahajan 2003:415).

It may be pointed out here that over a million of labour migrate out of Mahaboobnagar district for lack of employment opportunities on account of drought like situation. These labour called as *Palamur* labour are employed by middlemen in employment patterns akin to forms of bonded labour. This is because these labourers owing to extreme poverty are struck in a vicious circle of indebtedness (Ramana Murthy and Wendy Olsen, 2001).

Socially Discriminated Groups and Agriculture

Socially discriminated communities classified as scheduled castes and scheduled tribes constitute 25 per cent of total population in Mahaboobnagar and 28 per cent in Nalgonda district respectively. Of these, 23 per cent of scheduled castes and 49 per cent of scheduled tribes are engaged in cultivation as their main occupation while 60 per cent of scheduled castes and 32 per cent of scheduled tribes are employed as agricultural labour in Mahaboobnagar district. In Nalgonda district, 12 per cent of scheduled castes and 42 per cent of scheduled tribes are engaged in cultivation, while 64 per cent of scheduled castes and 42 per cent of scheduled tribes are employed as agricultural labour. Thus, agriculture has this significant social dimension of providing livelihood source for largest segments of these discriminated communities.

Dry Land Farming in Telangana

North and South Telangana put together, the percentage of irrigated area is 40.5 per cent. The percentage of irrigated area in South Telangana where the study is located is much worse. Only 33.5 per cent is irrigated area. Rest of the cultivation is rainfed (S. Subrahmanyam and P. Satya Sekhar, 2003:239). It is argued that poor yield of certain crops is linked to soil and climatic conditions (p.240).

The pressure on agriculture with reference to lack of proper irrigation facilities has been increasing over the years. It is observed that 'composition of irrigation has changed over the period especially in the recent years. The proportion of area under canal and tank irrigation has declined while the importance of well irrigation has gone up substantially. This may be attributable to the decline in public investment in agriculture in the recent years. These changes have environmental consequences. The increase in area under well irrigation coupled with the decline in tank irrigation invariably results in overexploitation of groundwater resulting in environmental problems such as desertification in fragile resource regions. Across, the regions, Telangana and Rayalaseema have experienced drastic shifts in the composition of irrigation' (V. Ratna Reddy, 2003:171-72).

It is stated that the groundwater is the single largest source of irrigation, covering, two-thirds of total cropped area under rainfed regions in Andhra Pradesh (*ibid.*, p.171). However, it is pointed out that the latest reports (based on official estimates) on groundwater situation are alarming. In the entire State, 336 Mandals are identified as problematic (*ibid.*, p.174).

Human Development Index and other Social Indicators

Of the 22 districts excluding Hyderabad city, on a three scale human development index of Education, Health and Standard of Living, Mahaboobnagar ranks 22nd while Nalgonda occupies the 13th rank (S. Subrahmanyam, 2003: 538).

It is pointed out that sex ratio in Andhra Pradesh increased by six points between 1991 and 2001 while South Telangana did not show any improvement. The literacy rates in Mahaboobnagar and Nalgonda were 44 per cent and 57 per cent respectively. However, of the total literates, female literacy in Mahaboobnagar was only 31 per cent and in Nalgonda only 44 per cent of females were literate. While 56 per cent of males and 69 per cent of males were literate in Mahaboobngar and Nalgonda districts respectively. Thus, position of women did not improve substantively in dry regions as it did in the wet regions (*ibid.*, p. 536).

Andhra Pradesh has the highest child labour. Percentage of child labour among children of the age group of 5-9 is 2 per cent and child labour amongst children of age group of 10-14 stands at 25 per cent in rural areas of Andhra Pradesh (S. Mahendra Dev and Vijay Mahajan, 2003:403). Further, it is pointed out that child mortality is positively correlated with poverty (S. Subrahmanyam, 2003:537).

Relevance of the Study

As already pointed out that bio-technologies are cheap, indigenous and easily accessible technologies that can contribute to increase the level of productivity. These technologies can be used by resource scarce regions and small and marginal farmers. Dry land agriculture amongst backward communities with low literacy has enormously benefited by introducing these technologies in countries like Kenya. It has also been pointed out that these technologies could have positive social implications in terms of having positive impact on gender relations and in reducing child labour.

It is quite evident from the statistics given in the Chapter 6 (Part II) that dependence on agriculture in the sites of research is very high. Further, there are large number of small and marginal farmers. It is also clear that these regions have poor literacy rates. Cultivation has been identified as having a significant social dimension in the fact that discriminated social communities are highly dependent on it for livelihoods. The sites of research chosen have been identified as lacking irrigation facilities and suffering desertification. Dry areas have shown poor human development indices, poor status of women and high instances of child labour.

Also, the Impact Assessment forms an integral part of any research oriented project. It becomes more relevant when it comes to assess the cost effectives of the technologies developed. It is quite mandatory that the return on investments made by the APNLBP is quantified to evaluate the net results of the projects delineated in terms of benefits derived for the farming community. It also becomes quite essential that the feedback from the farming community is obtained to further improve the efforts put forth by the Scientific community in developing a more meaningful and useful product of service. Thus, given this reality, this study proposes to assess the socio-economic impact to select bio-technologies developed for improving the livelihoods of dry land agricultural farmers.

3

Research Design and Methodology

Research Questions

The secondary data available from these areas suggests that there have been augmentation of incomes for farming households introducing these technologies. Given this backdrop, the present study proposes to address the following research questions:

1. What impact did the bio-technologies have on the small and marginal dry land farmers in pre- and post-biotechnology phases? How have discriminated social groups benefited from them?
2. Did the augmentation of incomes have an impact on nature of social relations and nature of employment patterns?
3. Have the bio-technologies been of any significance to gender relations in terms of greater participation ?
4. Have the bio-technologies and the employment generated by the farming activity based on bio-technologies prevented distress migration?

Objectives

Keeping in consideration the conceptual research questions that present study seeks an answer to the overall aim of the study is to examine the role of high technology, specifically biotechnology, in improving the socio-economic conditions of rural poor through sustainable livelihoods approach. However, the specific objectives of the study are:

- to document the ideation and execution mechanism of selected technology,
- to assess its social impact on the livelihoods of rural peasants,
- to examine the possibility of upscaling transferred technology,
- to suggest policy measures for sustainable livelihoods through biotechnology.

Research Approach

On the basis of overall aim and the nature of the study, it was decided that, at the conceptual level, it would draw from both the traditions of research approach, i.e., positivist and social constructivist. The positivist tradition stresses that the researcher is a disinterested, unbiased observer, who can produce objective truths about reality, a reality which straightforwardly exists out there to be observed. The social constructivist tradition challenge the view that a truly objective science is possible and researcher's standpoint affects what is seen at every level. So, in this tradition it is believed that reality is to some extent constructed by researcher's views of it.

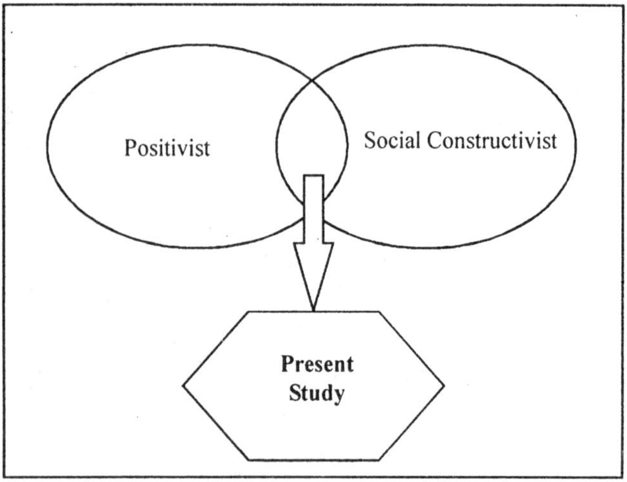

The study has taken some elements from positivist tradition as it involves natural sciences and is concerned with finding out the objective truths. At the same time study has also brought in the elements of social constructivist tradition, as it is people centred.

Based on this overall framework of research tradition, the study has followed a robust multi-pronged approach for ensuring its effectiveness in coming out with insights and conclusions.

Methodology

General

Based on the nature of research questions and the specific objectives, the study followed descriptive research design. The main aim of this

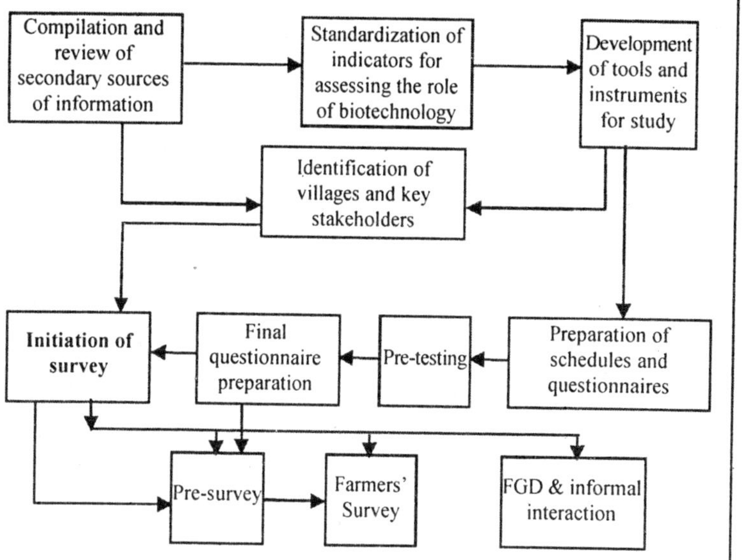

Schematic Representation of the Study

design is to present the situation and status of technology adoption at grassroots level. The general methodology of the proposed study included an intensive, participatory and on-the-field study and series of consultations at local, and district levels.

Sample Design

The sample design used was multi-stage sampling. At the highest level, the technology to be studied was selected in discussions with the BTU officials from among a host of technologies developed. At this level, *purposive sampling* was used in consultation with the officials for the selection of technology.

Selection of the Technology to be Studied

As there are a large number of technologies that have been developed and transferred for assisting the farmers with the help of APNLBP, a thorough discussion on the selection of technology for this study was made with the officials of APNLBP. After the discussions, it was felt that the present study should focus on "**Network Project on Agroforestry**" which is being implemented since 1999 and its second phase is nearing completion. This project was selected so that the study would bring out attitudinal changes that accrued for the implementation of the project. Also the study would try and focus on the community based organization which implemented this project and how much CBOs bring about increased social cohesion through an increment in the social capital of the perspective farmers.

Selection of the Study Area

Since the Network Project on Agroforestry in its first phase was implemented in the six villages of Mahboobnagar and Nalgonda, the study conduction was confined to these two districts only and that too in the villages where the technology has been transferred and adopted by the rural poor. The selected technology was implemented in the six villages of Nalgonda and Mahboobnagar. Among them, four villages were in the Mahboobnagar and two villages were in Nalgonda.

So, at this second stage, *stratified random sampling* was used and fifty per cent of the data points, i.e., villages were covered in proportion to that of the population. So, in all, at this level three villages were randomly selected. Out of these three villages, two villages were selected from Mahboobnagar district and one village was selected from Nalgonda district in order to adequately represent the visible patterns in the population.

Selection of Respondents

After deciding about the specific villages, the sampling frame for each specific village was taken from APNLBP. The sampling frame was the complete list of beneficiaries of the implemented technology. From the sampling frame, thirty-five villagers were randomly selected by means of lottery in order to remove any biasness in the selection of beneficiaries. In addition, from each selected village, fifteen non-beneficiaries were selected through *availability sampling*. Hence it can be said that the sampling design used was a judicious mix of probability and non-probability sampling techniques in order to adequately represent the population characteristics in the sample. The adequacy and representativeness of sample for the population was always kept into consideration in order that the findings and analysis made on the sample can be generalizable for whole of the population.

Data Collection Methods

Secondary Data Collection

The secondary information was collected from related publication, documents, research papers, partner NGO involved and organization involved in developing the technology. A major source of collection of secondary data was Biotechnology Unit of APNLBP. Apart from these, the secondary data was also collected from FAO, World Bank, UNDP. The secondary source will include review of related literatures and documents as described under SISCA approach.

Primary Data Collection

Following techniques were used to collect the primary data from different stakeholders of the project.

Survey of Principal Stakeholder

Based on the sample design, a survey of farmers who have adopted the technology was conducted to collect the information. Structured questionnaire, with a mix of open-ended as well as close-ended questions, was developed for this purpose. Also, the survey included those farmers who have not adopted the technology so as to get the cross-sectional view. In all, the data were collected from hundred respondents of the three villages who were part of the project and also the data was collected from fifty respondents who were not the part of the project.

Informal Discussions

Informal discussion was held during research/survey period with different stakeholders. These included beneficiary farmers, non-beneficiary farmers, the family members of the farmers, NGO personnel, local leaders, agricultural officers, NGO representatives, APNLBP staff and other staff of organisations involved in developing and extending the technology. This method included individual interaction in the form of in-depth interviews and group discussions. In due course of the study, three focus group discussions were held exclusively with the villagers of three selected villages for the present study.

Data Processing and Data Preparation

The quantitative data obtained from the field were tabulated and edited so as to check for any glaring discrepancy. All the field team members were specifically asked to make sure that the data collected by them are having minimum error. The examination, verification and editing of data were done at two levels. First at the field level, data were checked by the data collector and subsequently examined by the field team leader. Second at the broader level, data were again re-checked and verified by the principal investigator of the present study. Any discrepancy thus found out was removed again by going to the field and collecting the data again. Apart from minor discrepancies, if it was found that the quality of data captured was not up-to-the-mark and the team was not able to get proper responses on important variables even after going back to the field, those responses were not included in the final data preparation. So, after final data editing cleaning and processing, the sample size of the beneficiary was reduced to ninety whereas the sample size of the beneficiary was reduced to forty-five.

After this the data structure and data preparation were done. The format for the cleaned and edited data were prepared in the Statistical Package for Social Sciences 15.0. This was done so that SPSS 15.0 could be used for the data analysis.

The analysis of data focused around the measurement and quantification of role of technology in improving the lives of poor. A dual approach consisting of quantitative as well as qualitative dimensions was used in analyzing the data and information collected during the study. Univariate analysis, hypothesis testing, cross-tabulation, Chi-square test and Analysis of Variance were employed for gaining useful insights from the primary data thus collected.

4

AgNet Project

Andhra Pradesh Netherlands Biotechnology Programme (APNLBP) is one of the four country programmes supported by the Government of the Netherlands as a part of developmental cooperation. These country programmes are in operation in Columbia, Kenya, Zimbabwe and India. The Indian Programme focuses on Andhra Pradesh. The overall objective of the Programme is to improve quality of life of resource poor people through biotechnologies. AgNet Project refers to Network Project of Agroforestry which was conceived, funded and implemented by APNLBP with the help of a host of different institutions in the form of a network.

Background

The majority of rural poor, about 370 million of the poorest, live in areas that are resource-poor, highly heterogeneous and prone to high risks. Their agricultural systems are small scale, complex and diverse. The worst poverty is often located in arid or semi-arid zones, and in mountains and hills that are ecologically vulnerable (Conway, 1997). These areas are remote from services and roads and agricultural productivity is often low on a crop by crop basis, although total farm output can be significant. Such resource-poor farmers and their complex systems pose special research and developmental challenges and demand appropriate technologies (Netting, 1993).

Also, for decades, agricultural research, development and extension stood biased to the better-off, more progressive farmers (e.g. Chambers *et al.*, 1989; Röling, 1988). For the most part, resource-poor farmers gained very little from the processes of development and technology transfer of the Green Revolution (Pearse, 1980). Many analysts of the Green Revolution have pointed out that the new technologies were not scale-neutral. The farmers with the larger and better-endowed lands gained the most, whereas farmers with fewer resources often lost, and income disparities were often accentuated (Shiva, 1991). Not only were technologies inappropriate for poor farmers, but peasants were also excluded from access to credit, information, technical support and other services that would have helped them use and adapt these new inputs. It is widely acknowledged that, for instance, the Green Revolution in Asia resulted in a tremendous production increase of rice, but also in an increased gap between resourceful and marginal farmers.

High technologies developed often require investment to purchase inputs or equipment that the poor farmers cannot make. In other cases, technologies may not provide a solution to the real problems of marginal farmers who grow their crops under conditions totally different from those at experiment stations. Heterogeneity in the agro-eco-system, as experienced and even exploited by farmers in marginal areas, is considered a serious obstacle in scientific research, and therefore, often avoided (De Steenhuijsen Piters and Fresco, 1997).

The areas characterized by traditional agriculture remain poorly served by the transfer-of-technology approach, due to its bias in favour of modern scientific knowledge and its neglect of local participation and traditional knowledge (Lappe *et al.,* 1998). When poverty alleviation and upliftment of rural poor is one of the imminent priorities in front of the international community, it then becomes imperative that the energies should be directed in targeting and working with and under the conditions of the marginalized poor farmers. The historical challenge at present is, therefore, to re-focus efforts on marginalized farmers and agro-ecosystems and assume responsibility for the welfare of their agriculture. The private sector and advanced research institutions have little interest in targeting such groups as they do not represent significant markets.

In order to benefit the poor more directly, the overall approach must be applicable under the highly heterogeneous and diverse conditions in which small-holders live; it must be environmentally sustainable and based on the use of local and indigenous resources. The emphasis must be on improving whole farming systems at the field or watershed level rather than specific commodities. Technological generation must be demand-driven which means that research priorities must be based on the socio-economic and environmental needs and circumstances of resource-poor farmers (Blauert and Zadek, 1998).

The urgent need to combat rural poverty and to conserve and regenerate the deteriorated resource base of small farms requires an active search for new kinds of agricultural research and resource management strategies. NGOs have long argued that a sustainable agricultural development strategy that is environmentally enhancing must be based on agro-ecological principles and on a more participatory approach for technology development and dissemination (Altieri *et al.,* 1998). Focused attention to the linkages between agriculture and sustainability will help greatly in solving the problems of poverty, food insecurity and environmental degradation.

To be of benefit to the rural poor, agricultural research and development should operate on the basis of a "bottom-up" approach, using and building upon the resources already available: local people, their knowledge and their autochthonous natural resources. It must also seriously take into consideration, through participatory approaches, the needs, aspirations and circumstances of small-holders (Richards, 1995).

For years several NGOs in the developing world have been promoting agro-ecologically-based NRM approaches. Agro-ecology provides a methodological framework for understanding the nature of farming systems and the principles by which they function. It is the science that provides ecological principles for the design and management of sustainable and resource-conserving agricultural systems—offering several advantages for the development of farmer-friendly technologies. First, agro-ecology relies on indigenous farming knowledge and selected modern technologies to manage diversity, incorporate biological principles and resources into farming systems, and intensify agricultural production. Second, it offers the only practical way to restore agricultural lands that have been degraded by conventional agronomic practices. Third, it provides for an environmentally sound and affordable way for small-holders to intensify production in marginal areas. Finally, it has the potential to reverse the anti-peasant bias of strategies that emphasize purchased inputs as opposed to the assets that small farmers already possess, such as their low opportunity costs of labour.

Ecological concepts are utilized to favour natural processes and biological interactions that optimize synergies so that diversified farms are able to sponsor their own soil fertility, crop protection and productivity. By assembling crops, animals, trees, soils and other factors in spatial/temporal diversified schemes, several processes are optimized. Such processes are crucial in determining the sustainability of agricultural systems (Altieri, 1995). Agro-ecology takes greater advantage of natural processes and beneficial on-farm interactions in order to reduce off-farm input use and to improve the efficiency of farming systems. Technologies emphasized tend to enhance the functional biodiversity of agro-ecosystems as well as the conservation of existing on-farm resources. One of the robust agro-ecological approaches is agro-forestry systems.

Ideation

The ideation and subsequent research and development for such a sustainable approach towards the upliftment of poor with the help of advancements in high technology comprises co-creative processes to identify the problems, generate new information and innovations, consolidate them with adequate existing farming practice, and then translate them into learning objectives and activities for enhanced farmer performance. These processes are likely to be highly iterative and synergistic. Participatory research targeting the needs of poor farmers should begin with collaborative identification and analysis of problems, needs and opportunities, in an attempt to gain an understanding of the broad agro-ecological and socio-economic context. This includes the identification of already existing alternatives to solve the problem(s), which may need to be tested under different conditions, and should eventually be consolidated with innovations. The problem identification phase should lead to the participatory priority-setting and formulation of the overall project goals and specific research objectives. The final output is a prioritized research agenda.

Keeping this in mind APNLBP ideated about initiating a participatory project which would address the needs of small and marginal farmers of Mahboobnagar and Nalgonda following an integrated agro-ecological approach of agro-forestry. To this regard, a one-day workshop was held on 29th December, 1997 which was attended by 29 participants from voluntary organizations and Government departments of Andhra Pradesh. Based on the discussions and summary of findings of this workshop, BPC, the highest decision making body of APNLBP, recommended for having an external consultancy on agro-forestry. After the feasibility study report which was submitted by the external consultant in September 1998, BPC gave the formal go-ahead for the agro-forestry project research agenda and it was formally named as *Network Project on Agro-forestry*.

Once the research agenda was set, innovation development followed. This phase included both a basic and an applied research component. Farmers' involvement in innovation development is particularly desirable at the level of applied research. Their role may vary from "analysts and evaluators" (Fano *et al.*, 1996) validating existing technologies to "research collaborators" determining and testing treatments in their own fields (Ashby *et al.*, 1995; Braun and Van de Fliert, 1997). So, in order to make the whole project fully participatory in nature, Rapid Resource Appraisal was done. Based on the results of the Rapid Resource Appraisal, constraints were identified and analyses were made. Likewise, developmental goals and objectives, as well as strategies to solve such constraints, were formulated, as shown below.

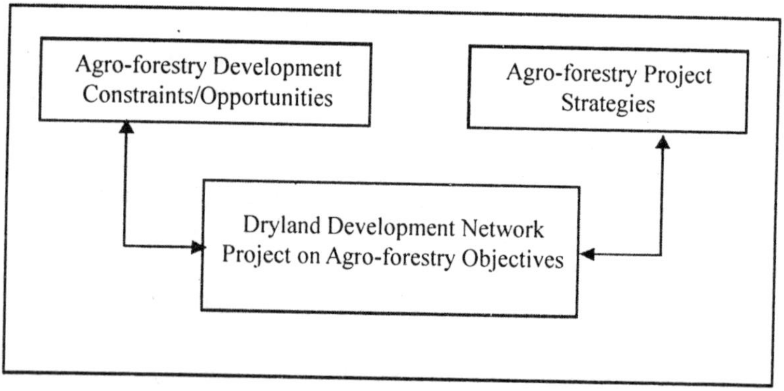

Fig. 4.1: Schematic Diagram of the Relationship among Constraints, Objectives and Strategies for Dryland Development by Network Project on Agro-forestry

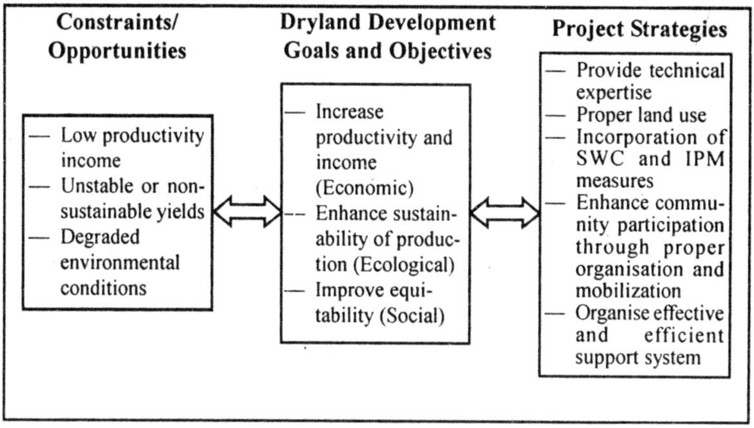

Fig. 4.2: Rapid Resource Appraisal Analysis in the Project Area of Network Project on Agroforestry

So it was agreed that broadly Network Project on Agro-forestry has conceptualized with a view that the farmers should get some kind of minimum assured income even under severe drought conditions and to reduce the migrations. Furthermore, it was formulated that the ultimate goal of the project is sustainable development of dryland agriculture in Drought Prone Areas. Specifically it may be said to include:

(i) Continuous improvement of productivity consistent with growing demands.
(ii) Control of land degradation and protection of environment.
(iii) Increased income of farmers and thus improved quality of their life.

After the formulation of objectives and overall goals the project area was selected.

Selection of Project Area

The six project sites are spread in Nalgonda and Mahaboobnagar districts. The project areas are typically drought prone areas with harsh climatic conditions. Rainfall is around 600 mm per year. In Mahaboobnagar and Nalgonda districts, about 30 per cent of arable land remains fallow every year as arable farming in such situations is not viable due to the continuous process of land degradation. This was done on the basis of Rapid Resource Appraisal as mentioned earlier.

"Development" (within the context of research and development) is defined here as the translation and validation of innovation development outputs in relation to the agro-ecological, socio-economic and cultural conditions in target areas. The development process should not end with applied research, which is often considered the final step of research mandates. Applied research should be followed up by deliberate attention to training and development. Experience has shown that linear, top-down research and extension, as practiced in conventional technology transfer models, often failed because of inappropriate technology and/or inadequate "packaging" of the messages (Röling, 1988). Moreover, consistency is needed between the nature of the innovation and that of the extension approach and methods applied to convey the innovation to farmers (Röling and Van de Fliert, 1997). Therefore, to ensure consistency, we should not only look at the innovations *per se*, but also define the capacities that practitioners need to implement them as well as the requirements for the support system (input supply, markets, etc.). This leads to an analysis and definitions of what a change in agricultural practice effected by the developed innovations implies for the farmers.

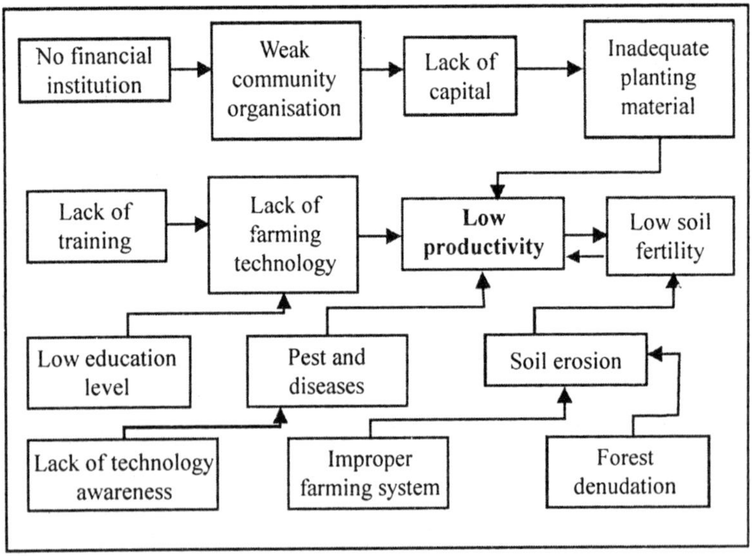

Fig. 4.3: Network Analysis of the Problems Faced in the Proposed Area for Network Project on Agro-forestry

Keeping these views in consideration, the following model was adopted before final execution and implementation of the project in order to dovetail the project objectives and the needs and constraints faced by the intended beneficiaries. It was also decided that during execution phase this model would be the guiding factor in prioritizing the activities so that it truly addresses the needs for which it was conceptualised and meant.

Execution and Implementation

Extension and implementation encompass the phases when efforts are made—either in a formal or a non-formal setting—to share the innovation with larger groups of farmers who then test, evaluate and incorporate (or reject) them in their farming practices. Changing farming practices should ultimately lead to substantive impact.

Extension—defined here as a function of disseminating an innovation to a wider audience—is not usually considered part of the mandate of research institutions (Fano *et al.*, 1996). Therefore, suitable mechanisms and partners must be found to perform this function. To ensure that potential partners can carry out extension work efficiently, scientists can play an important role, contributing both technical and methodological skills. These skills may be complemented by those of GO or NGO extension workers, who have a comparative advantage as communicators at the village level.

In many developing countries, extension services lack the human resource capacity—in terms of both quantity and quality of staff—to reach a critical mass of their target audience effectively (e.g., Röling, 1988). Much of the information obtained by farmers is disseminated by other farmers, either directly by sharing experiences or indirectly through demonstrations of sample field practices and the resulting effects. Recent experience with IPM training in several Asian countries has shown the positive impact of involving farmers as trainers, and of enhancing farmer networks in order to support farmer-to-farmer dissemination deliberately (Eveleens *et al.*, 1996; Braun, 1997). Farmer facilitators must be selected with care and given additional training on facilitation methods. Training programmes must also address farmer interaction and horizontal communication requirements from start and during the planning stage.

The major actors in the implementation realm are, of course, the farmers. Farmers decide whether or not to implement, adapt or reject an innovation. Enhanced knowledge and skills obtained through training, contact with fellow farmers or any other form of learning are catalysts for change in farming practices. While much research has been devoted to studying the process of adopting innovation (Rogers, 1995), in terms of sustainable agriculture, adapting innovation—to farm-specific conditions—is considered a more valuable output particularly under the marginal conditions of poor farmers. The ability to adapt *guidelines* rather than follow a standard recommendation is evidence of farmers' enhanced capacity to experiment, analyse, evaluate and, finally, solve many of their own problems without having to depend upon external advice. Response mechanisms, however, are critical in this realm because farmers often receive contradictory messages from other sources (e.g., promotional campaigns by commercial companies selling alternate inputs), which could lead to confusion. Questions arising during implementation need to be addressed by trainers, whose role includes supporting the adjustment process and helping bridge communications between farmers and researchers. When farmers' capacities and practices change, tangible effects at the farm level can be expected. These may include yield increase, reduction of expenditure, or more balanced ratio of pests to natural enemies in the field. When such changes occur on a larger scale, an even broader impact can be expected, such as the improvement of rural people's livelihoods and/or a healthier environment. If initial outputs prove beneficial to farm families, they will most likely be disseminated further, contributing to a general increase in the knowledge base of the farming community.

In the light of above discussions and the concerns raised therein, a robust model of planning, execution and implementation was put in place. This can be conceptually shown as in the adjoining diagram.

As shown in the diagram, the responsibility of implementation of the Network Project was invested in CRIDA due to its technical expertise. Central Research Institute for Dryland Agriculture (CRIDA) is one of the premier institutes of the Indian Council of Agricultural Research (ICAR) catering to the location of specific needs of dryland farming in different agro-ecological situations through its all-India network. The network is comprised 22 Dryland Farming Research

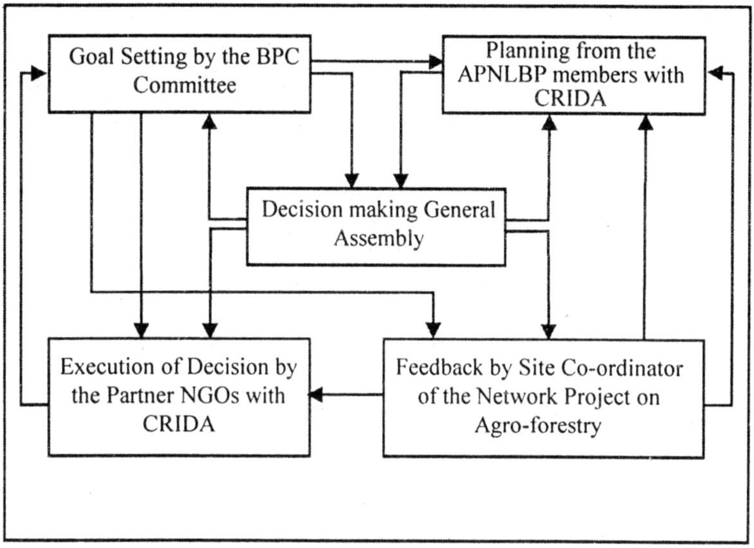

Fig. 4.4: Model of Planning for Dryland Development by Network Project on Agro-forestry

Stations under All India Co-ordinated Research Project for Dryland Agriculture (AICRPDA) and 24 Agromet Research Stations under All India Co-ordinated Research Project on Agrometeorolgy (AICRPAM). Besides AICRPDA and AICRPAM, strategic research in the frontier areas is being conducted at two research farms of CRIDA located in Ranga Reddy District of Andhra Pradesh. The role of CRIDA is specifically mentioned as below in the implementation of this project:

- Fine tuning of the proposals submitted by NGOs with respect to identification of appropriate technologies,
- Guidance for efficient implementation of the projects and on-farm participatory technology development,
- Human resource development and monitoring and evaluation.

In addition to CRIDA which was supposed to implement the project and provide technical expertise, six NGOs were identified which would help in assisting the extension, execution and implementation of the project. The NGOs participating in the programme were selected on the basis of the willingness of the NGO to participate in the programme. In addition, the following criteria were applied in their selection process for better facilitation of project activities.

- Location of the NGOs in the districts,
- Capacity of the NGO to take up the programme in terms of available staff, technical background etc.,
- Earlier work done on similar projects,
- Rapport of the NGO with the target community.

On the basis of these criteria, the following six NGOs were selected to assist in the project. These were:

1. People's Action for Creative Education (PEACE),
2. Sri Aurobindo Institute of Rural Development (SAIRD),
3. Research in Environment, Education and Development Society (REEDS),
4. Villages in Partnership (VIP),
5. Society for Development of Drought Prone Area (SDDPA),
6. Vasavya Society for Rural Development (VASORD).

Role of NGOs

The selected NGOs were expected to play the following role in the programme:

- Selection and motivation of farmers,
- Implementation of various project components under the guidance of CRIDA,
- Organising training programmes and exposure visits for farmers,
- Organising farmers' groups,
- Participating in interactive group meetings,
- Sourcing and obtaining the planting material under the guidance of CRIDA,
- Nursery raising with farmers participation,
- Submission of progress and financial reports to IPE through CRIDA.

Selection of Participants

Since the project aims at participation of poor and marginal farmers, it was decided to include those farmers who belong to the dryland area, having interest in the project, are small and marginal holders and their main source of income is through agriculture and allied activities. There was no negative criterion held by the project. In addition, special care was taken to make the selection process bias-free in terms of gender and caste.

Formation of Community-based Organization

As a prerequisite of implementation of project, the selected beneficiaries were encouraged and motivated to form community-based organization. This was done in order to facilitate the implementation of the project and to ensure the sustainability and viability of the initiative after the completion of project. Each of the selected NGOs were specifically made responsible to nurture and develop these CBOs and motivate the farmers to discuss and settle different issues that might arise in due course of project implementation through the CBOs. For this, it was decided to hold monthly meetings of the CBOs in the respective villages. In addition, at a higher level, meetings/discussions would be held with representatives of CBOs, NGOs, CRIDA team and BTU officials for the smooth execution of the project.

Farmers Training

CRIDA is entrusted with the responsibilities of organising training programmes for both farmer's and NGOs. One-day training programmes were organised at CRIDA farm. Interactive meetings organised periodically also had served to reinforce the training inputs. In addition, during the monitoring visits to individual NGOs, CRIDA team had provided the site-specific information to farmers. Training courses on agro-forestry technology were also conducted to teach the farmers the techniques of dryland farming. Many cross-farm visits to other successful agro-forestry sites were made to enable the farmers to see for themselves and to interact with their fellow farmers.

Exposure Visits

Exposure visits to other agro-forestry programmes have reinforced the farmers understanding of the need for agro-forestry. These exposure visits had also helped the NGO personnel in better conceptualisation of the agro-forestry models required for these areas. This has greatly helped in shaping the present agro-forestry models. In addition to motivating the farmers and NGOs, these visits have helped them to visualize the potentials of tree-based farming systems in dryland situations. Most of the project participants have been taken on these visits.

Monitoring and Evaluation

The monitoring and evaluation realm forms a maze, overlapping with and collating the other two realms. Researchers must observe and measure what happens during training and implementation, and must relate and/or recycle the information back to the research and development realm for further adjustment or impact assessment. Systematic monitoring and evaluation of projects assures the capacity to make adjustments before it is too late, to learn from experiences and to justify the research investment. Rapid feedback is critical when farmers are presented with new variables (for example, a new variety, a new technology, or a more complex, integrated innovative approach). In participatory projects, monitoring and evaluation should be planned and implemented in conjunction with the farmers. Farmers should particularly be involved in defining indicators for evaluation, and in analyzing evaluation results. In the case of sustainable agriculture, evaluation indicators should always relate to the objectives and expected outputs of each phase. Within this context, well-defined indicators usually focus as much or more on people and the environment than on technology and economics (Van de Fliert, 1998). Monitoring and evaluation of clearly defined indicators should generate valuable feedback for adjusting current project methodology, improving future research and development, and providing examples for other projects.

In order to be able to justify the research and development investment, the monitoring and evaluation system should be designed to analyze the outputs in relation to the objectives set for each specific phase. This is depicted by the horizontal links in Fig 4.1, where the expected outputs of the activities and elements in the extension and implementation realm relate directly to the objectives of the activities in the research and extension realm at the same horizontal level. After the evaluation exercise the objectives set forth can be either said to have been achieved or not achieved. These horizontal links clarify the idea that in order to have positive impact, research and development teams should seek mechanisms for incorporating extension and implementation requirements when setting their objectives for research and development.

Drawing upon the discussion above, a robust monitoring and evaluation mechanism was put in place. At the basic level, CRIDA was made responsible for monitoring and evaluation due to its technical expertise and important role in the implementation. In addition, an official from BTU was made specific in-charge to see whether the project is being implemented on the stated guidelines and within the broad framework and objectives of APNLBP or not. In addition, at the higher level external teams were appointed at regular intervals to evaluate the project from an outsiders' perspective. Overall, BPC also took up regular discussions in its meetings regarding overall monitoring and evaluation of the project and the NGOs involved therein.

Farmers' Participation

Participation of farmers in planning and implementation has greatly enhanced the effectiveness of research efforts in addressing the needs of the poor (e.g. Chambers *et al.*, 1989; Scoones and Thompson, 1994; Haverkort *et al.*, 1991). A second important focus for poverty-oriented work is an integrative nature of methodologies, to ensure congruity of analytic, disciplinary research with farmers' holistic perception to their farm management, and to needs and opportunities for improvement (Van de Fliert and Braun, 1999).

Most developmental programmes that placed the interests of small-scale farmers high on their agenda fell short in their expectations as they failed to seriously address popular participation. The few existing examples of generation and diffusion of "farmer friendly" technologies suggest that full participation of farmers is essential to the development and dissemination of sustainable agriculture methods and technologies. In such cases horizontal and equitable interaction among actors replaces top-down relations, and promoted initiatives are responsive to farmer needs and ideas (Thrupp, 1996).

The existing farmer-to-farmer networks and methods of communication have proven invaluable in the spreading of ideas and innovations. In turn, these participatory arrangements strengthen and empower local farmer and community organizations, and further learning and adoption of alternatives (Selener *et al.*, 1997).

Farmer participatory research projects aim at achieving tangible impact by encouraging farmer involvement. Achievement of impact is an ambitious goal, requiring both qualitative change (e.g., effect on farmer capacities, practices, collective actions, or support systems) and quantitative change (i.e. on a considerably large scale). Careful planning at each project stage to ensure (1) appropriate problem definition and establishment of objectives, (2) successful generation of applicable information or innovations, and (3) appropriate development and use of dissemination mechanisms and their effective implementation, can contribute to the achievement of this kind of impact. The level of farmers' participation should be determined by their interests and capacities and by the nature of the work. In such collaboration, farmers provide the holistic perspective, explain what works and what needs to be improved in the current system, set the evaluation criteria for innovations in accordance with their objectives for farming, and test possible innovations. Technical scientists contribute new technological information that may provide options for improvement, and methodologies for testing the various options. Social scientists help identify the constraints and opportunities in the support system, seek ways to reduce the constraints and enhance the opportunities, and translate technological innovations into farmer learning objectives.

The above discussions were kept in mind right from the ideation stage of the project up to the monitoring and evaluation. In fact, special care was taken to ensure that beneficiaries are completely involved in any and every activity of the project. Due to this reason, only a robust ideation mechanism keeping beneficiaries central to it and respecting their knowledge base was adhered to. Similarly in every stage of execution and implementation, participation of beneficiaries was ensured. Since the project villages selected happen to be the regular villages of the concerned NGOs, there is good rapport of the NGO staff with the villagers. The participants readily responded to the suggestions of the NGOs. This has also resulted in the establishment of good farmers' CBOs who are taking keen interest in the project activities. In turn, this has generated a great deal of awareness among them and their knowledge base has now melded with the technical inputs provided by the project.

Further Steps

An integrative, participatory approach to agricultural research and development does in itself not guarantee that impact will be achieved among those who need it most. In certain cultural settings, the implicit dominance of the better-off in a community may cause problems: Often, the wealthier community members are perceived to have a higher social status, preventing the use of a truly participatory approach, especially when the methodology involves community-level meetings. Therefore, prior understanding of local power and communication structures are important when designing project methodology. Targeting the needs of the poor implies giving more weight to their opinions. To compensate for this, a variety of methods should be used during needs assessment activities to ensure input from all layers of the community. In addition, various "tricks", as described in textbooks on participatory methodology (e.g., Van Veldhuizen *et al.*, 1997) can be used to prevent biases.

A critical element of methodology design and preparation is site and participant selection. The sites and participants should have characteristic representatives of all major target group(s). For example, if smallholder farmers under marginal conditions are targeted, research must be conducted with such farmers in marginal areas, even if this entails difficult logistics (e.g., transport to remote areas). The project should preferably be implemented in relatively remote communities where no other current or recent projects have been conducted.

Another important component in the process of participatory developmental programme is to ensure that there is continuous feedback among the various actors in the process. Joint reflection on the findings of research activities ensures better consistency across components, and mutual compatibility with new components and adequate existing practices. This can be done, for instance, by organizing regular evaluation and planning workshops at the project sites, involving representatives from all stakeholder groups.

Encouraged by the responses from the farmers and the robustness of the mechanism put in place, the project has been extended to its second phase. But in this phase, care needs to be taken on the specific points discussed above like site-specificity, constant feedback mechanism, prior understanding of local power and communication structures, site and participant selection among other things.

5

Analysis and Findings—Part I

As indicated earlier, an exhaustive and robust methodology for the collection of data for the present study was adopted. This included in-depth personal interviews of the beneficiaries along with non-beneficiaries. This was done in order to have a cross-sectional view and a holistic perspective about the project. In addition to in-depth interviews, Focus Group Discussions with the community and interviews with other stakeholders in the project were also conducted. For the study the data was gathered from the three selected villages where the first phase of project had already been completed. From these three selected villages, a sample of thirty beneficiaries from each of these villages was taken for in-depth interviews. Also, an interaction in the form of in-depth interviews was made with a sample of 15 non-beneficiaries from each of the selected villages. So, in all, the data were collected from 135 respondents from the three selected villages.

Sample Characteristics

A brief discussion on the characteristics of the sample is desirable as a prelude to the analysis and findings of the study. This would add to the meaningful discussion on the findings and would add robustness to it.

Gender-wise Description

On the whole, out of a sample of 90 beneficiaries who consented to be the respondents of the present study, ninety-one per cent were male and nine per cent were female representing the pattern of the population of beneficiaries. Coming to village-specific description,

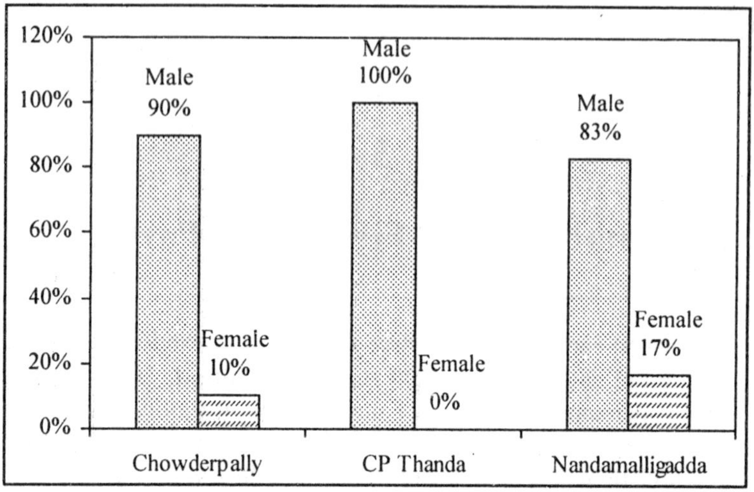

C.P. Thanda had no representation of female beneficiaries whereas Nandamalligadda had a seventeen per cent representation of female beneficiaries. It is worth mentioning that the pattern discerned here in the sample bears a strong resemblance of the characteristics exhibited by the beneficiary population of the project under study. So, it can be said that the first phase of the project may not have put in specific efforts in the direction of addressing the gender bias which

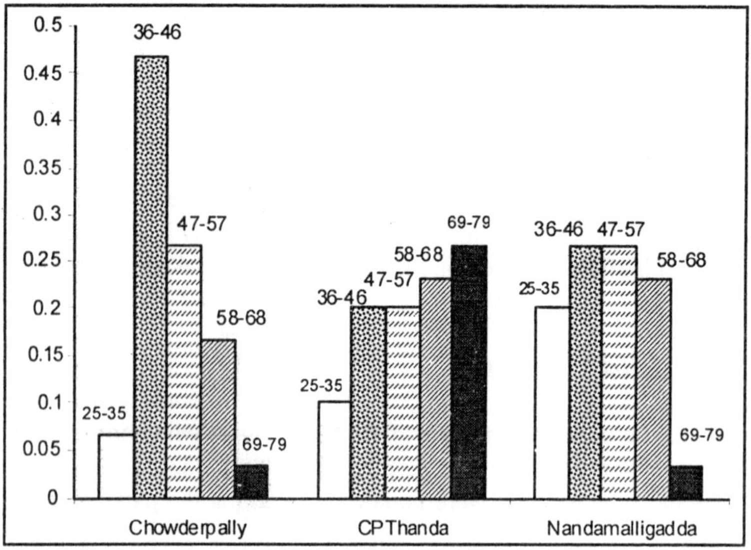

is a prerequisite for the sustenance of initiatives of this nature. Addressing gender gap always results in the sustainability of the initiatives as evidenced from the outcomes of numerous projects such as Grameen Bank.

Age-wise Description

The mean age of the respondents who agreed to participate in the present study was 50 years. This reveals the fact that the beneficiaries were mature enough and were having considerable experience. At the same time at the village level there is a difference in the mean age of the respondents. The respondents of the Nandmalligadda and Chowderpally have mean ages of 47 and 48 years respectively, whereas the respondents of C.P. Thanda have mean age of 55 years which is considerably higher than the other two villages. This may be due to the fact that half of the respondents of C.P. Thanda fall in the age group of 58 and above as depicted in the adjoining graph.

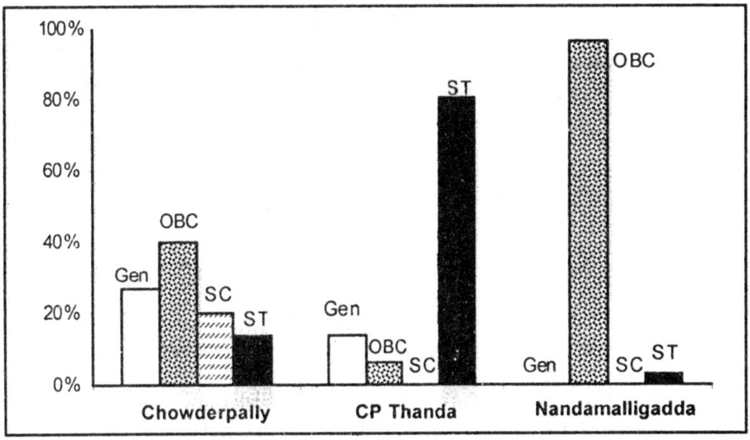

Caste-wise Description

The sample has a dominance of OBCs and STs as is evident from the graph. On the whole, OBCs have 48 per cent representation and STs have 32 per cent representation in the sample. Village-wise, Chowderpally represented all the castes in fair proportions resulting in greater heterogeneity among the beneficiaries whereas, Nandamalligadda did not have representation from General caste and

SCs. This may have resulted in better homogeneity among the beneficiaries of the project in Nandamalligadda. In terms of homogeneity, C.P. Thanda would be in the midst of these two buttressing the overall trend in the population evident in the respective villages. For the beneficiary population of the first phase of this project, the pattern is more or less similar, stressing the fact that the project targeted weaker sections of the society.

Education-Level-wise Description

As may be expected, being a study of rural focus, the sample had 52 per cent respondents who were literate. Probing a bit further, it

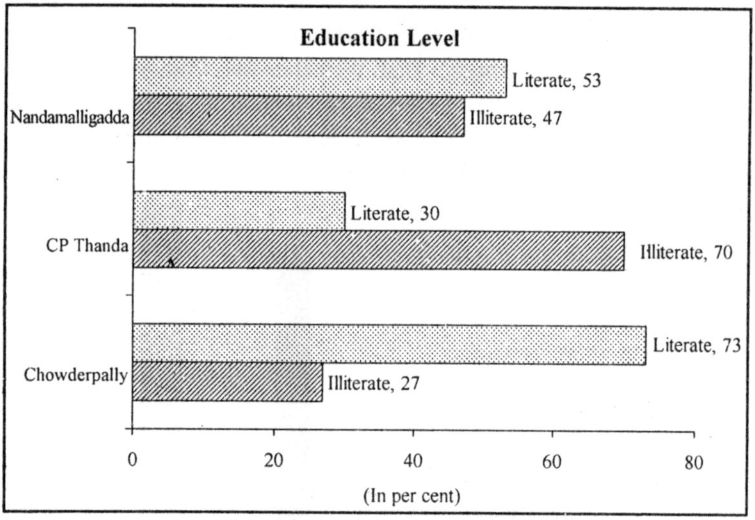

can be discerned that illiteracy is predominant in C.P. Thanda which is a tribal hamlet, whereas in Chowderpally, 73 per cent of the respondents were literate. In Nandamalligadda, 53 per cent of the respondents were literate.

Occupation-wise Description

As can be expected, the sample had 90 per cent of respondents whose primary occupation was farming. Another 6 per cent of the

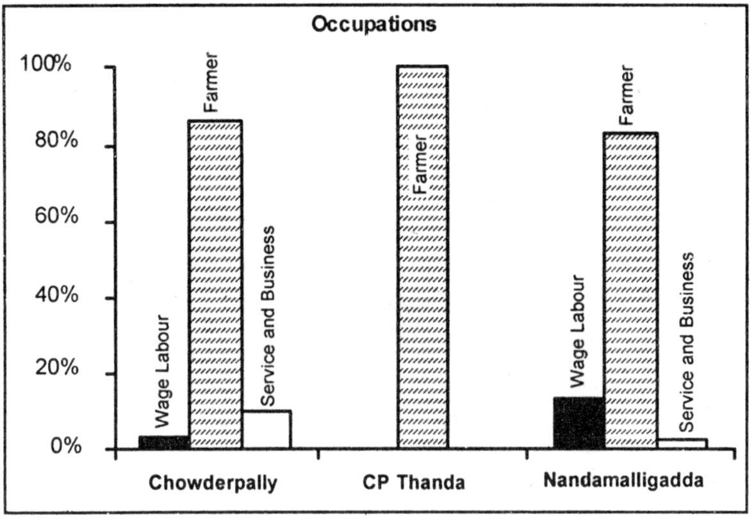

respondents were wage labourers and only 2 per cent each of the respondents had their primary occupation as service or petty business. But these respondents were again involved in agriculture as their secondary occupation. As can be seen from here, the respondents did not have a diverse portfolio of income generation and practically all of them were dependent on agriculture for their livelihood. This is particularly true in case of C.P. Thanda where all the respondents gain their income from farming. Chowderpally relatively has a more diversified portfolio of income generation as can be discerned from the adjoining graph.

Description on Household Type

As can be discerned from the adjoining graph, C.P. Thanda and Chowderpally has a very high percentage of respondents residing in *Pucca* house (more than 80 per cent). In comparison to this, in Nandamalligadda, the proportion of respondents residing in *Pucca* house is slightly more than half. From this it can be inferred that, among the respondents hailing from the less privileged sections of the society, economic disparity among them is more pronounced in Chowderpally and C.P. Thanda as compared to that of respondents from Nandamalligadda. Moreover, it may also happen that the

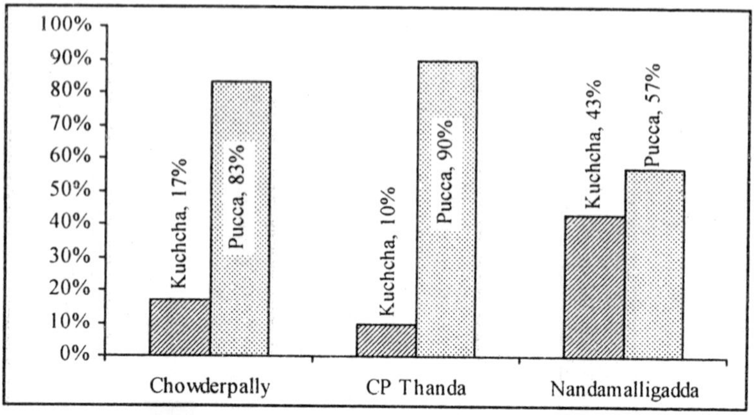

respondents from C.P. Thanda and Chowderpally may not participate in the project as there is less incentive for them to do the same.

Agro-forestry Awareness

Pre-project Awareness about Agro-forestry

As can be seen from adjoining graph, majority of the respondents, i.e., 96 per cent of them were completely unaware about agro-forestry

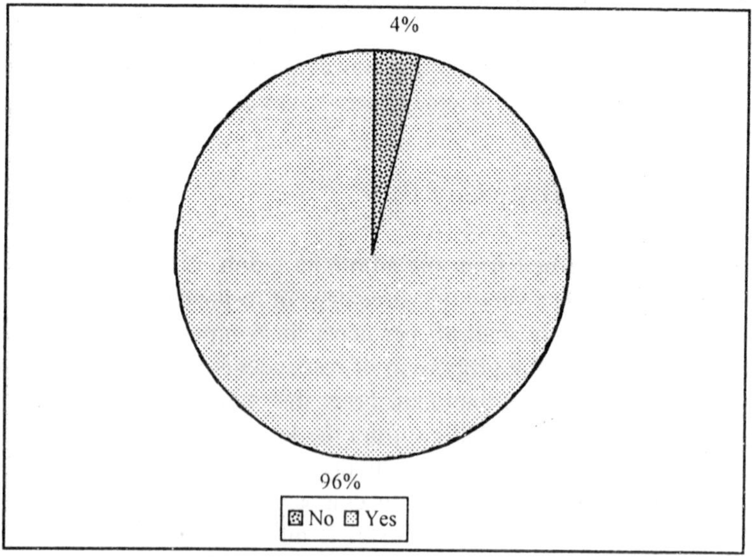

before commencement of this project. Only a few (4 per cent) of the respondents had a bit of knowledge regarding agro-forestry. In the light of this, it can be said that the selection of project area for the implementation of agro-forestry project with a stated aim of extending the technological was well suited.

A village specific analysis reveals that all of the selected villages for the study where the project was implemented had overwhelming majority of the respondents being completely unaware about the concept of agro-forestry before the commencement of the project. The adjoining graph shows that in all the three villages the respondents were not having any idea about agro-forestry before this project. Among these three villages, Chowderpally had 10 per cent respondents who knew a bit about agro-forestry. Therefore, we can infer that the level of awareness among these respondents is substantially low, which becomes even more challenging for the NGOs to implement and sustain this project.

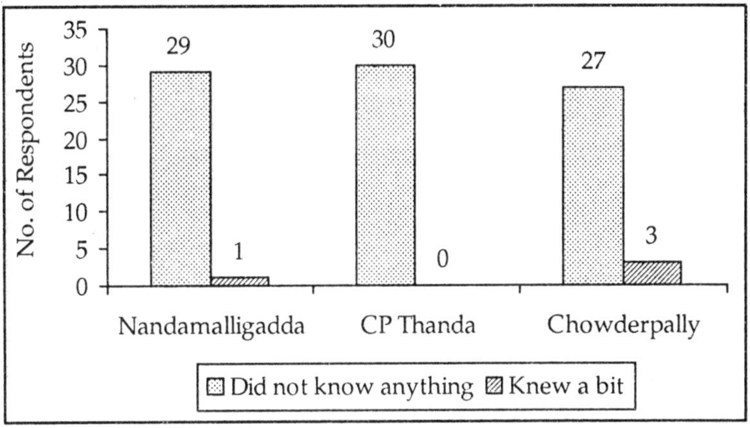

Pre-project Awareness among the Family Members

It can be observed from the adjoining graph that most all the respondents, i.e. 99 per cent said not even a single member of their family was aware about agro-forestry. It shows difficulty of motivating NGOs and scientists to convince people for implementation. It is also observed that a majority of the family members of respondents in all the three villages were unaware about agro-forestry. In

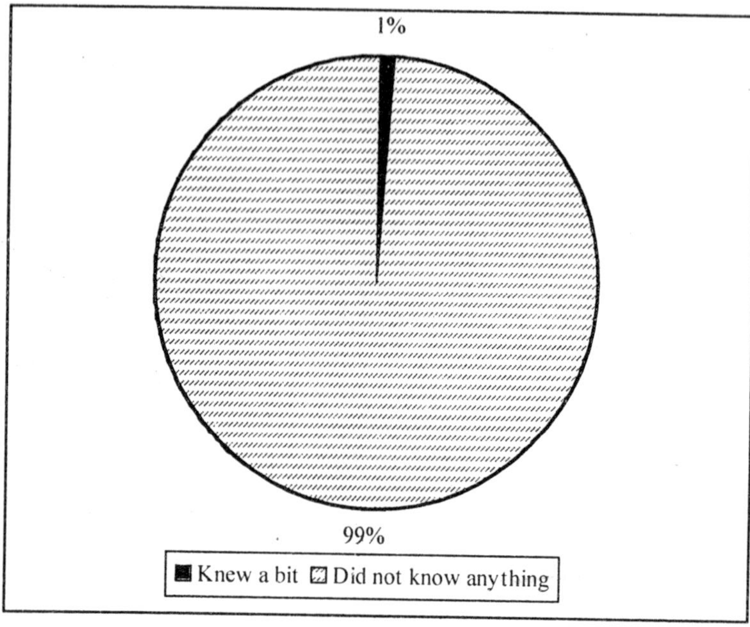

C.P. Thanda village not a single-family member of respondents was aware about agro-forestry. Maximum numbers of awared people were in Chowderpally, i.e. 10 per cent of total respondents in that village. Therefore, we can infer that in all the three villages, unawareness is common.

Source of Information about Agro-forestry

It can be observed from the adjoining graph that majority of the respondents came to know about agro-forestry project either by association with projects, i.e. 73 per cent of the respondents or from other practicing farmers, i.e. 24 per cent of respondents. Maximum number of the respondents got information about agro-forestry from those people who are directly involved with agro-forestry projects.

It can be observed from the adjoining graph that in all the three villages a majority of the respondents came to know about agro-forestry after associating with the project. In Chowderpally and C.P. Thanda a good proportion of respondents came to know about agro-forestry from other practicing farmers, i.e. 40 per cent in C.P. Thanda

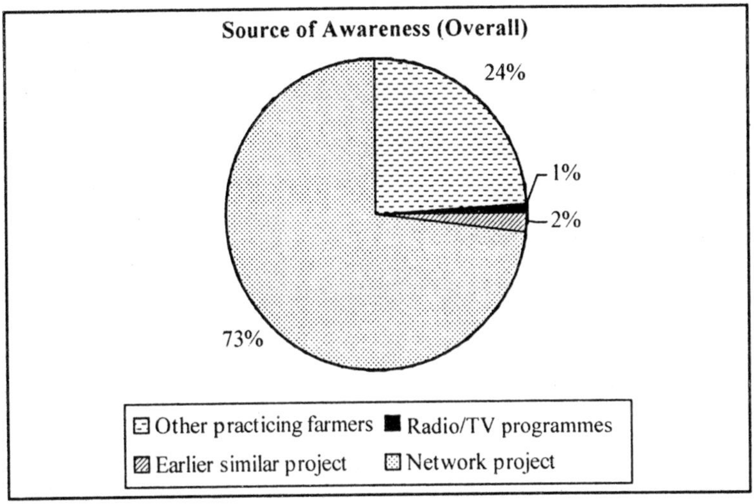

and 26.66 per cent in Chowderpally. We can infer from here that a majority of the respondents it was completely new and unknown thing for them.

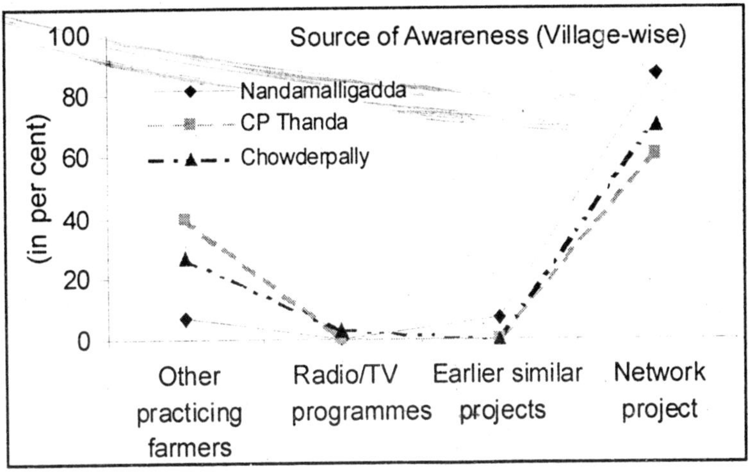

Reason of Association with the Project

As can be seen from adjoining graph, half of the respondents said that the direction of NGO was a reason for associating with agro-forestry project. Second major cause for associating with this project

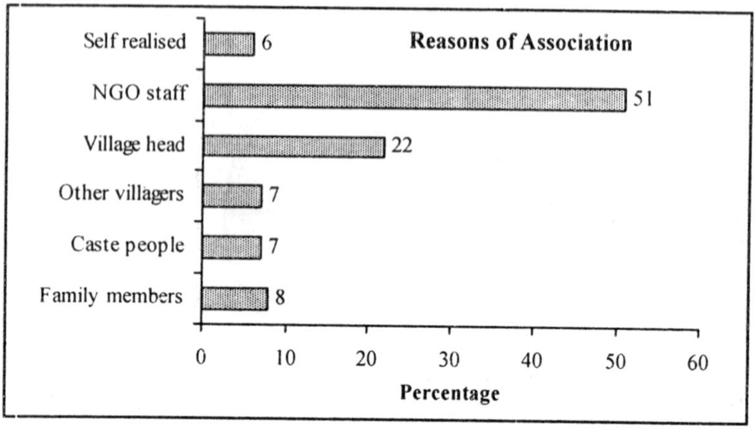

is direction by village head and other functionaries. A significant number of respondents, i.e. almost one-fourth of total respondents, were influenced by other factors like family members, caste people, and association of other villagers and self-realization of benefit. It can be inferred that 50 per cent of the respondents joined the project as per the advice of the NGO that reveals the effectiveness of NGO in persuading the people to join the project and spread the benefits they can derive from the project.

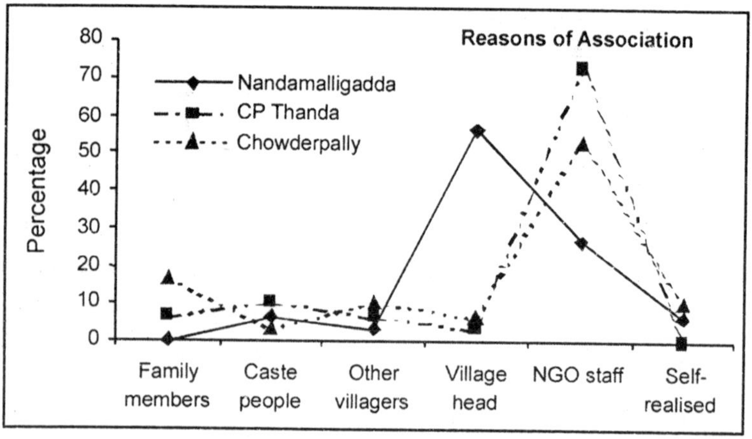

It can be seen from the adjoining graph that respondents in all three villages were highly influenced by direction of NGOs and directions of village head and other functionaries. Nandamalligadda is the only village where maximum of the respondents were influenced by village head and other functionaries, i.e. 57 per cent of total respondents. Except Nandamalligadda, in rest of the two villages, a majority of the respondents were influenced by direction of respondents, i.e. 73 per cent in C.P. Thanda and 54 per cent in Chowderpally.

Expectation from the Project

It can be seen from the chart that 44 per cent of the respondents associated with the project expecting that it will help in increasing their income; whereas 33 per cent of the respondents chose to join the project expecting that the project will help them in increasing the productivity of the land. So it can be inferred that a majority of respondents chose to associate with the project as they felt that it would increase their income and productivity of the land.

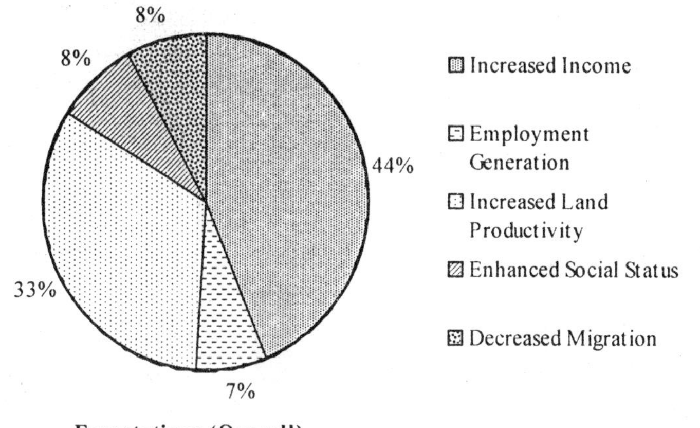

Expectations (Overall)

As can be seen from the above graph, in which two out of three villages' respondents expected that the project would increase their income. Nandamalligadda is the only village where respondents expected that project would enhance productivity of their land. We can infer that most of the respondents in all the three villages were expecting either increase in income or in the productivity of their land.

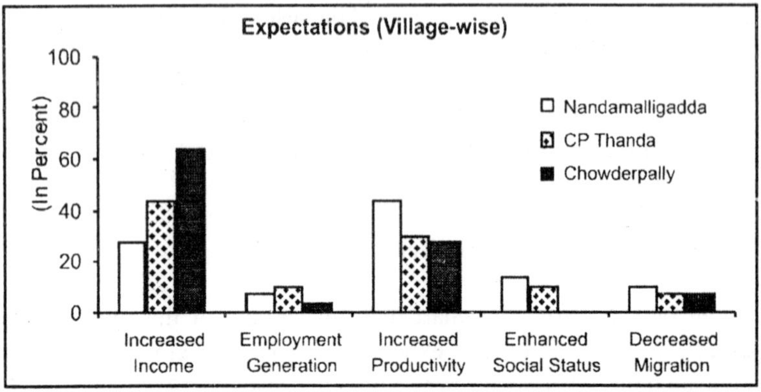

Along with the people who expected that their income would increase, majority are from the literate side. Out of the total sample, maximum number of respondents are illiterate whose expectations are highly different from respondent to respondent, i.e. 44 per cent are expecting an increase in income, 30.23 per cent are expecting an

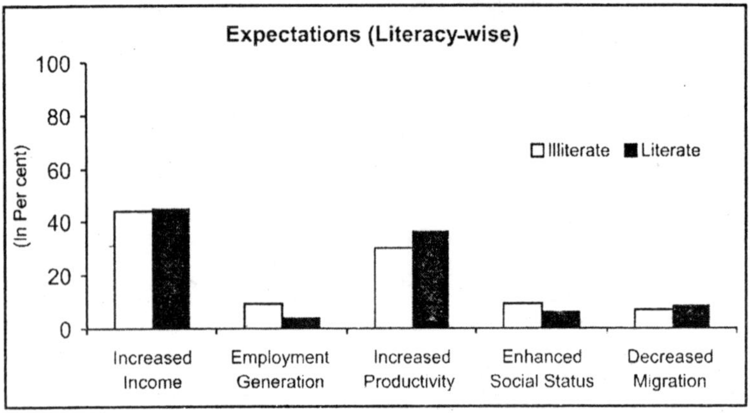

increase in productivity of land and rest one-fourth of illiterate respondents are expecting the project would increase employment opportunity, social status and stop migration.

Awareness of Respondents Regarding Agro-forestry after the Project

It can be seen from the chart that 6 per cent of the respondents feel that their awareness about agro-forestry has increased by one-fold. Whereas 17 per cent of the respondents feel that their awareness about agro-forestry has increased by two-fold, 29 per cent and 27 per cent of the respondents feel that their awareness about agro-forestry has increased by three-fold and mani-fold respectively. Only a minuscule

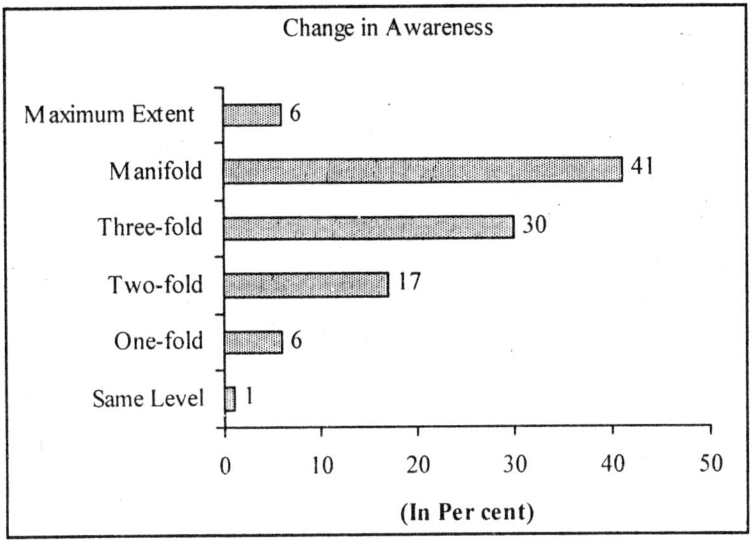

part of the respondents feel that their awareness about agro-forestry has remained the same. It can be inferred that the project has been effective in increasing the awareness about agro-forestry among the respondents.

Adoption

Support from the NGO

While trying to gauge the extent the NGO/ Scientists found solutions to the problems the respondents faced, the analysis revealed that,

46 per cent of them replied they received help most of the times from the Scientists/NGO. It can also be observed that 20 per cent of the respondents were of the view that they received help several times from the scientists. A small part of the respondents, i.e. about 1 per cent have said that they have not received any help from the Scientists/NGO. It can be perceived from the above that respondents had various degrees of satisfaction with respect to the help they received from the Scientists/NGO.

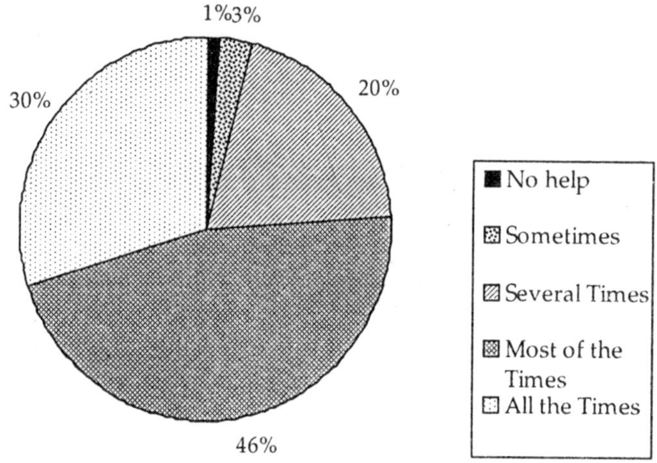

Assistance from NGO/Scientist (Overall)

Severity of Constraints

Dearth of Technical Information

It can be witnessed from the adjoining chart that 45 per cent of the respondents said that they had not at all faced any dearth of technical knowledge whereas 42 per cent of respondents were of the view that they had faced less dearth of technical information. On the other hand it can be observed that a miniscule part of the respondents responded that they had faced a severe dearth of technical information. It can be perceived that the Scientists/NGO were successful in disseminating required technical information to the needy participants.

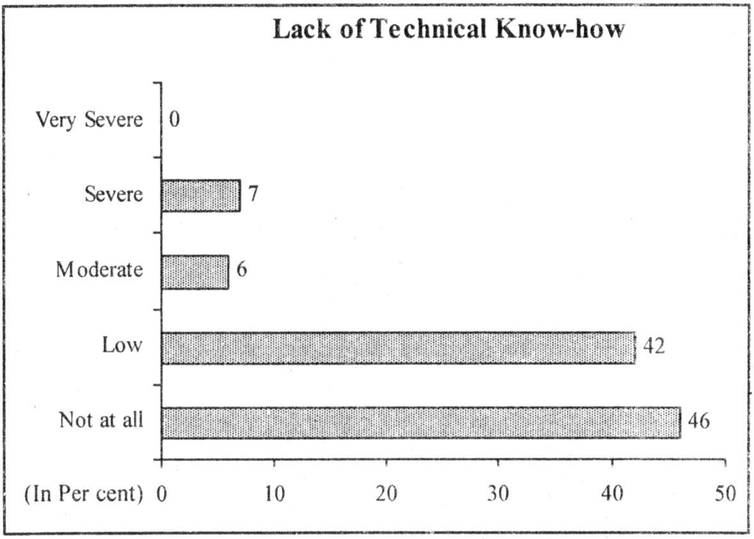

Supply of Seeds

It can be inferred from the chart that a majority of the respondents replied they had not faced shortage of seed supply whereas 29 per cent of the respondents had faced a low shortage. On the other hand a minuscule part of the respondents had severe shortage of seed supply.

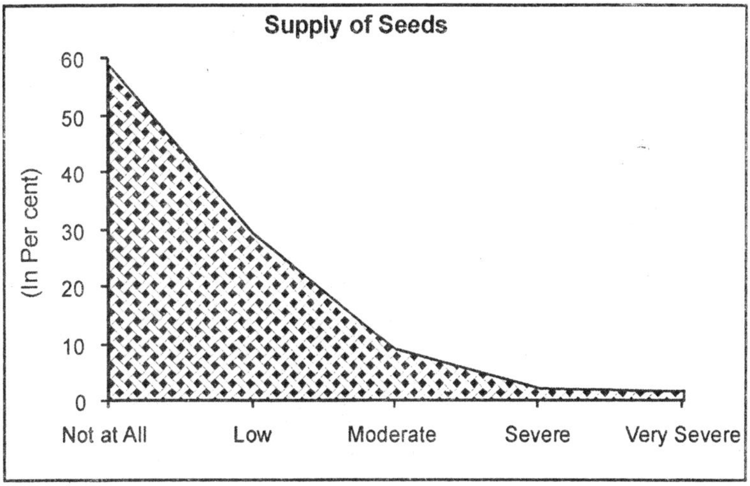

So it can be viewed from the above data that the NGO was effective in supplying seeds when required by the participants.

Supply of Fertilizers

It can be seen from the adjacent chart that a majority of the respondents had no problem with the supply of fertilizers whereas 22 per cent of the respondents were of the view that they had faced a low supply of fertilizers. On the other hand, a minuscule part of the respondents

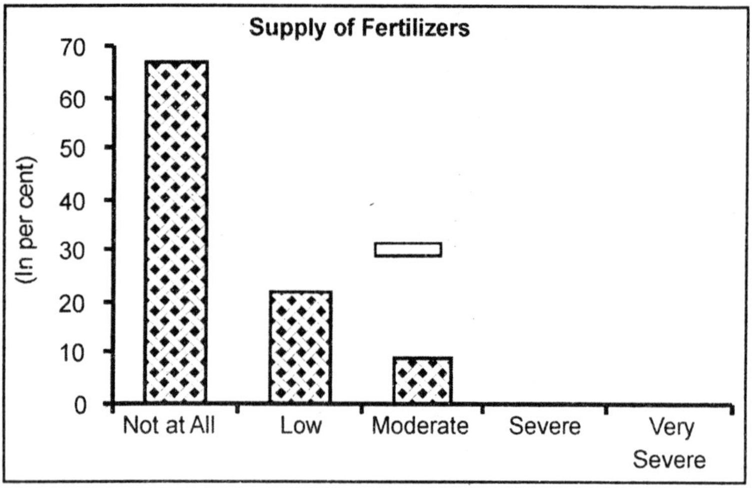

had severe shortage in supply of fertilizers. So it can be viewed from the above data that the NGO was effective in supplying fertilizers when required to the participants.

Other Constraints

It can be seen from the adjacent chart that 39 per cent of the respondents had no problem with the availability of other resources whereas 43 per cent of the respondents had some problem. On the other hand a minuscule part of the respondents had a severe problem with the availability of other resources. So it can be viewed from the above data that the NGO was effective in providing other resources when required by the participants.

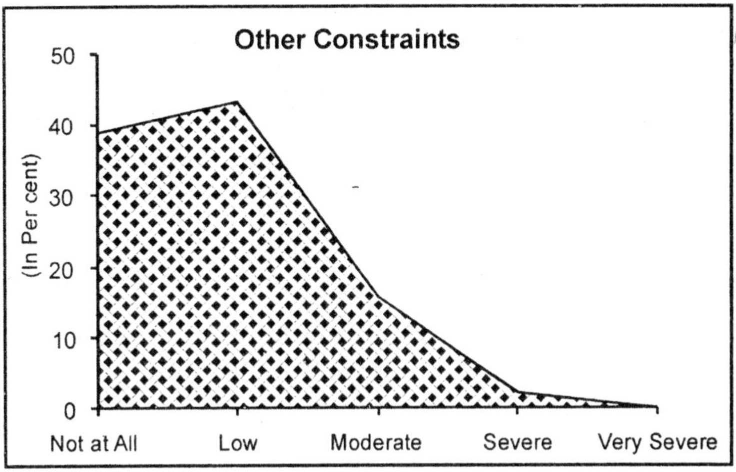

Overall, on a five point rating scale, where five corresponded to "very severe" and one corresponded to "not at all severe"; all the constraints were measured by the respondents. Lack of technical information received a score of 1.733 showing that degree of severity was low. Similarly, supply of seed received a score of 1.577; supply of fertilizers received a score of 1.466 and other constraints received a score of 1.811. This reveals that the adoption process did not face any severe constraint in the implementation of the project.

CBOs and Processes

Formation of Community-based Organization

For the successful implementation of the project and ensuring the sustainability of the same, formation of community-based organization is vitally important. Keeping this in mind, an attempt was made to see the extent of formation of CBOs in the study area selected. The villagers of Nandamalligadda, C.P. Thanda, and Chowderpally said that they formed as an organization with the help of the implementing NGO. And also it can be seen that the respondents' response in all the three different villages is similar and positive thus, we can say that the NGO has helped a lot to the respondents in forming an organization.

Decision-Making Committee

The analysis reveals that 97% were of the view that the community-based organization had a committee which takes decision on behalf of members of the committee. Just a minuscule part of the respondents said that no such committee was formed which takes decisions on behalf of its members. It can be perceived that community-based

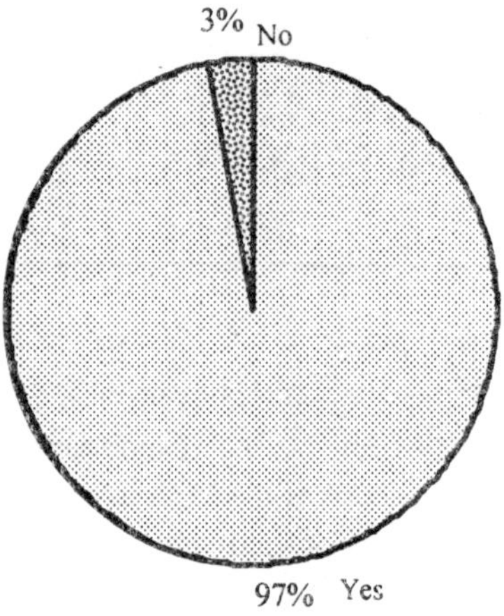

Presence of Decision-making Committee (Overall)

organization has be formed and it actively works on behalf of the members in taking decisions and in implementing them which definitely will increase the chances of the sustainability of the organization as well as the efforts of NGO in inculcating the methods of agro-forestry among the villagers.

Formation of Committee for Decision-Making

As can be observed from the adjoining figure, almost all of the respondents in all three different villages have shown a positive reaction. Only in C.P. Thanda and Chowderpally there is a very small number of respondents who had given negative feedback.

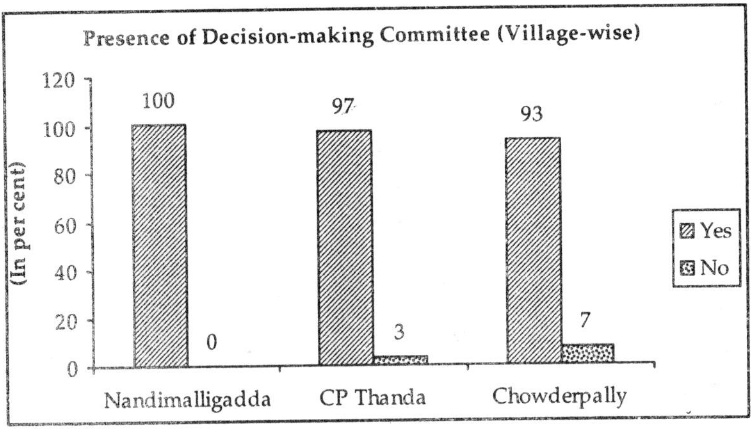

Methods Used to Select Committee Members

As can be observed from the adjoining figure, the majority of the respondents have said that members chosen for the committee in the village is through selection. On the other hand 26 per cent of the respondents said that members chosen in the village is through Election. And a very few respondents have said that committee members are chosen from consensus among the people.

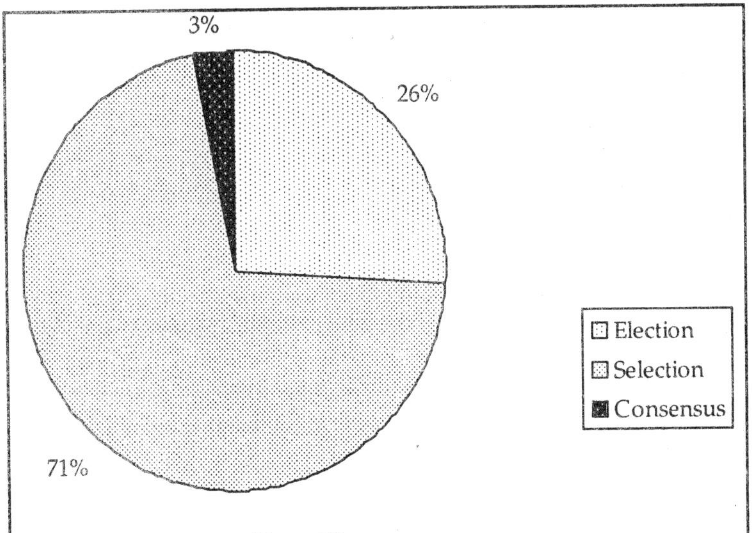

Selection Process (Overall)

As can be observed from the adjoining figure, that majority of the respondents have said the members chosen for the committee in C.P. Thanda and Chowderpally village is through selection, and the

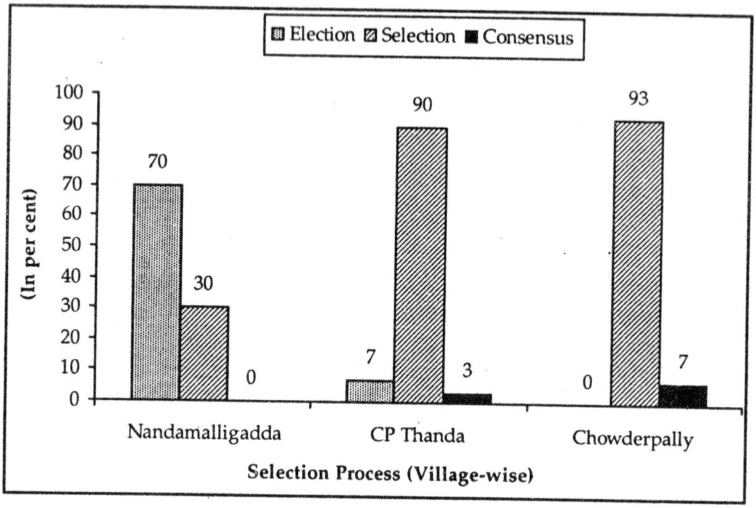

majority of the respondents have said that members chosen in Nandamalligadda village is through Election. And in all the three different villages only a few respondents have said that committee members are chosen through consensus among the people.

Formation of Rules and Regulations

It can be seen from the adjoining chart that a majority of respondents were of the view that rules and regulations were framed or were in place for effective implementation of the project. On the other hand, 14 per cent of the respondents said that there were no such rules in place which could facilitate effective implementation of the project, whereas the remaining 14 per cent of the respondents were not aware of such rules and regulations.

It can be perceived that the NGO has been successful in formation of community-based organization and in establishing rules and regulations for effective implementation of the project.

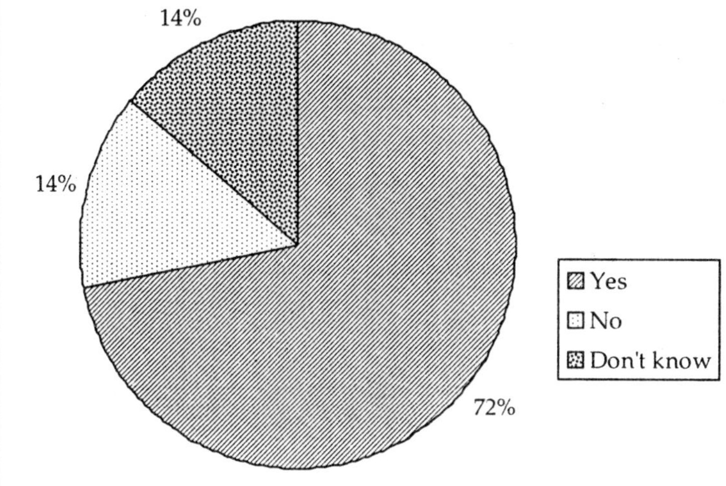

Elective Implementation (Overall)

As it can be seen from the adjoining figure, a majority of the respondents in all three different villages have given positive response for organization-formulated rules and regulations for effective implementation of the project in Nandamalligadda and Chowderpally,

Effective Implementation (Village-wise)

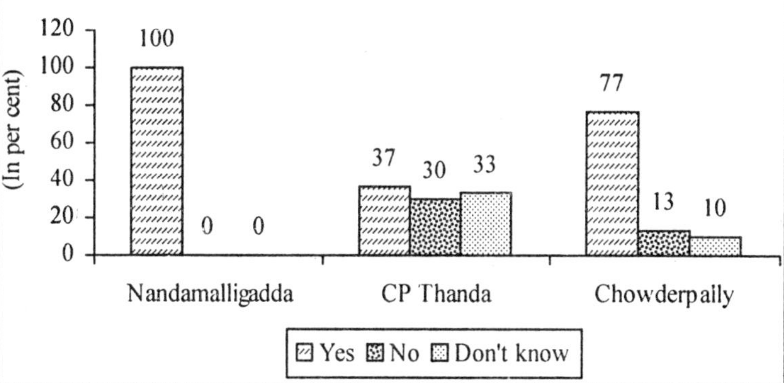

there is almost negligible respondents who said that organization did not formulate any rules, neither they did not know about it. But, in C.P. Thanda the response was quite close in all the three criteria.

Frequency of Conflicts Related to Group Activities

It can be seen from the adjoining chart that 48 per cent of the respondents said that conflicts relating to group activities arise frequently. Whereas 39 per cent of the respondents were of the view

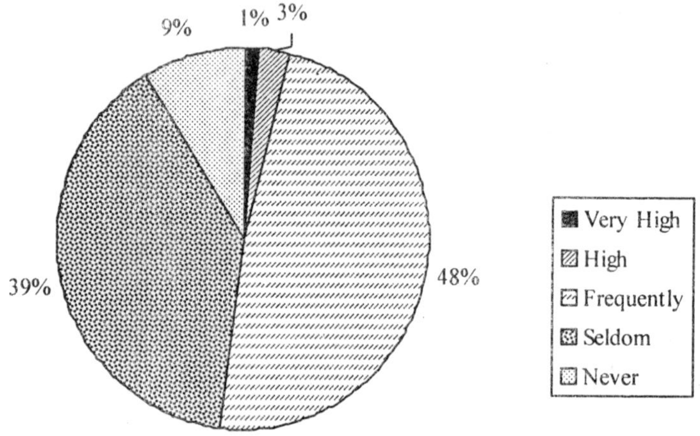

Frequency of Conflicts (Overall)

that such conflicts arise in almost all the meetings, 1 per cent of the respondents said that such conflicts were very high. Thus it can be inferred that NGOs have been successful in establishing community-based organization and in implementing rules and regulations for effective implementation of the project, but were not effective in avoiding the frequent occurrences of conflicts.

As can be seen from the adjoining figure, majority of respondents in Nandamalligadda said seldom about the conflicts related to group matter and similar number of respondents said they frequently conflict or never. In C.P. Thanda village, majority of respondents said they have conflicts frequently, and only a few respondents said they seldom have conflicts or very high conflicts in every meeting. In Chowderpally, majority of respondents have said they have conflict related to group matter frequently, and a good number of respondents gave response as seldom, and almost negligible respondent gave feedback towards never, every meeting, very high.

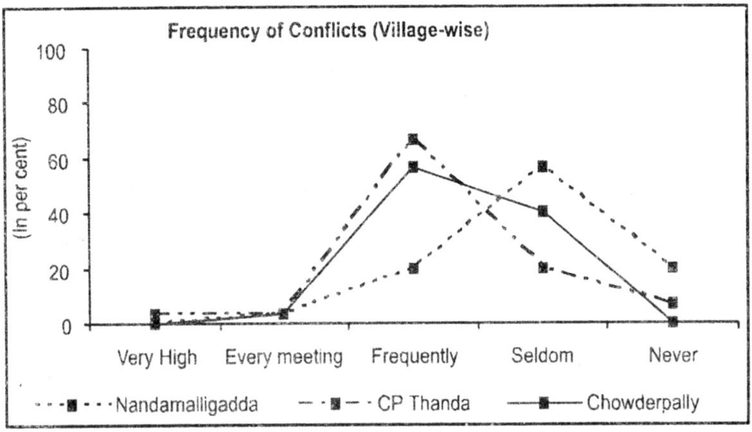

Conflict Resolution Mechanism

It can be seen from the adjoining chart that 48 per cent of the respondents said that leader of the committee resolved the conflict whereas 16 per cent of the respondents were of the view that some external body resolved the conflict. On the other hand, 28 per cent of the respondents had no idea as to how the conflicts are resolved in the committee.

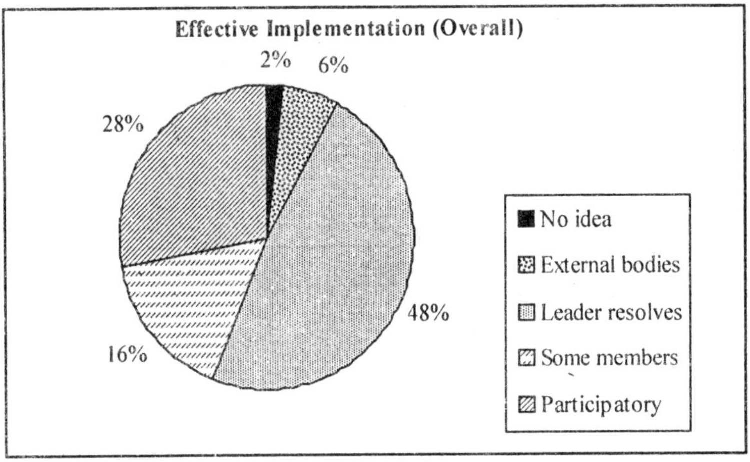

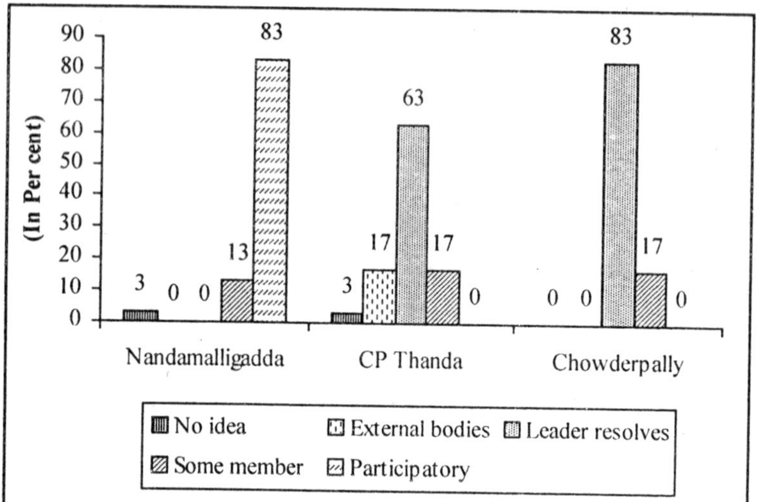

Effective Implementation (Village-wise)

As can be seen from the adjoining figure, majority of respondents of Nandamalligadda village have said that they resolve conflicts mainly through participatory manner, and very few respondents said that some members always resolve the conflicts. In C.P. Thanda village, majority of respondents have said that the conflicts are resolved mainly through the Leader, and similar number of respondents gave response as external bodies or other some members would resolve. In Chowderpally village, majority of respondents have said that conflicts are resolved by Leader, and very few respondents gave response as some members do it always.

Frequency of Conflict Resolution

It can be witnessed from the chart that a majority of the respondents said that the conflicts related to group activity were resolved in the same meeting in which they arose; whereas 16 per cent of the respondents were of the view that such conflicts were resolved in the succeeding meeting or in the next meeting. On the other hand, 26 per cent of the respondents said that such conflicts were resolved as and when time was found and a miniscule part of the respondents

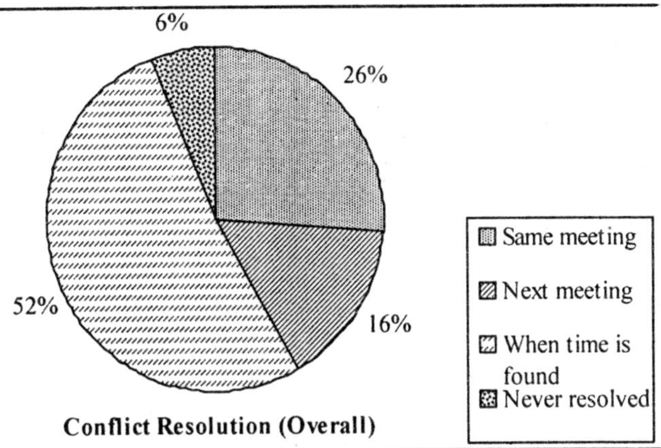

Conflict Resolution (Overall)

i.e. 6 per cent of them responded that such conflicts are never resolved. It can be seen that the community was effective in resolving the conflicts related to group activities.

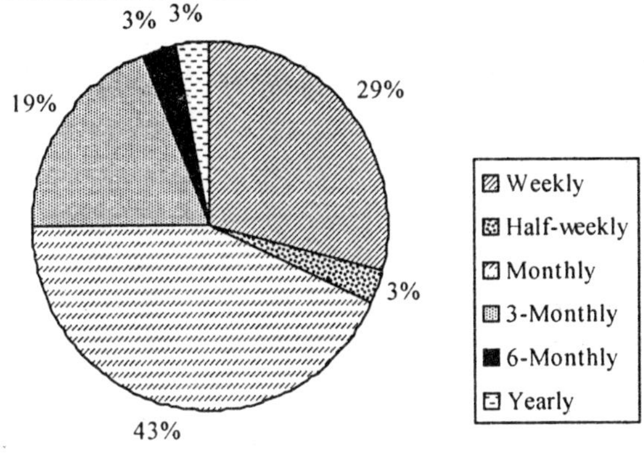

Meetings (Overall)

As can be seen from the adjoining figure, in Nandamalligadda village the majority of respondents have said that they resolve conflicts in the same meeting, and some of the respondents said that they would resolve the conflict in the next meeting. In C.P. Thanda village, the respondents gave response for all options (i.e. same meeting, next meeting, when time found, never resolved).In Chowderpally,

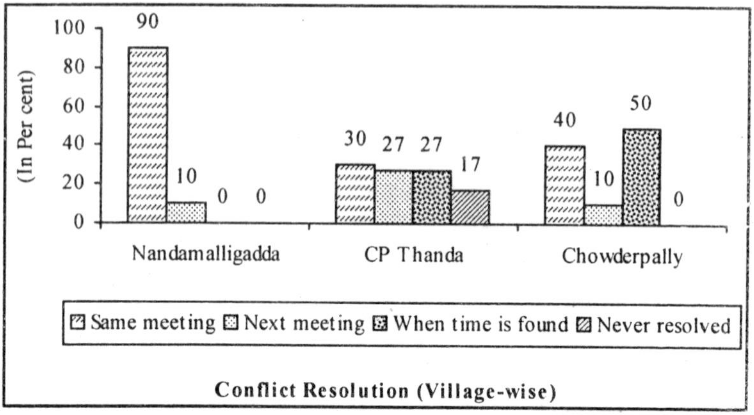

majorityof the respondents have said that they resolve conflicts when they find time, and also a good number of respondents have said that they resolve conflicts in the same meeting itself and very less number of respondents said that they resolve the conflict in the next meeting and no respondent said that they are never left resolved.

Frequency of Meetings Held to Discuss Issues Related to the Project

About 92 per cent of the respondents were of the view that members met for discussing the issues related to the project. Only a small part of the respondents, i.e. 7 per cent of them were of the view that the members do not meet to discuss about the problems and remaining 1 per cent of them were unaware about such meetings. Of the 92 per cent of the respondents, 53 per cent of them told that the meetings are held once in a month, 27 per cent of the remaining told that meetings are held once in every three months. It can be perceived that NGO has been successful in ensuring that members meet regularly to resolve the conflicts related to the group and also to solve the problems of its members.

As it can be figured out from the adjoining graph, the respondents said that the frequency of meeting in the Nandamalligadda village is more in weekly and very less in half-monthly and monthly, and there

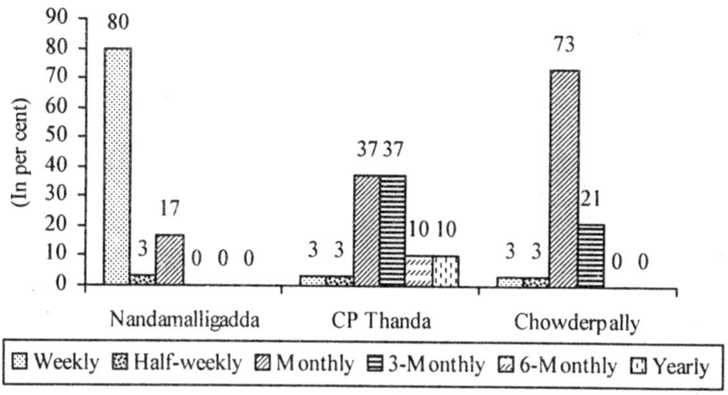

Meetings (Village-wise)

is no frequency of meeting in 3 months, 6-months, and in the year. In C.P. Thanda village, the respondents said that the frequency of meeting is very similar in monthly and in 3-months basis, whereas the frequency of meeting is less in week, half-month, half-yearly and yearly manner. In Chowderpally village, the respondents have said that the frequency of meetings is more in months, and less in week and half-month, there is no frequency of meeting in 6-months, yearly.

Networking Relationships

Horizontal Relationships

It can be inferred from the chart that a majority of respondents said that there exists a moderate level of horizontal relationship between the members of the organization, whereas 17 per cent of the respondents said that there exists a high level of horizontal relationship between the members of the organization. It can be perceived that the NGO has been successful in establishing mutual understanding and relationship between the members of the organization.

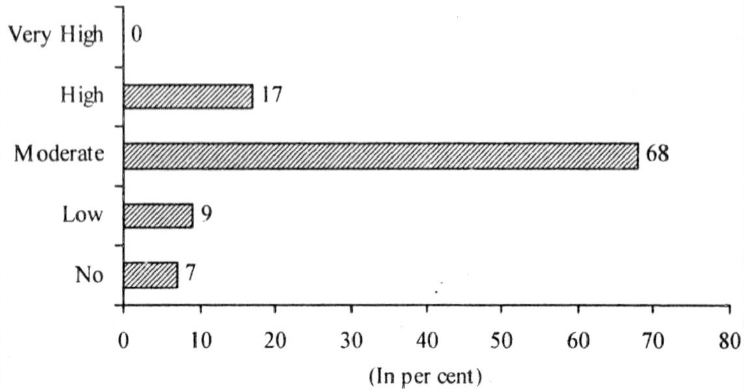

Existence of Horizontal Relationship (Overall)

Vertical Relationships

It can be witnessed from the chart that a majority of respondents said that there exists a moderate level of vertical relationship between the members of the organization. Whereas 17 per cent of the respondents said that there exists a high level of vertical relationship between the members of the organization. It can be perceived that the NGO has been successful in establishing horizontal and vertical relationships between the members of the organization.

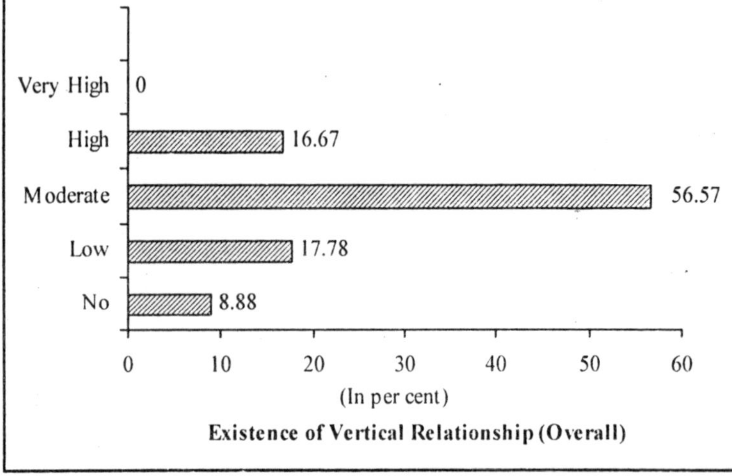

Existence of Vertical Relationship (Overall)

Impact of CBOs on Networking

On the issue of formation of CBOs, 100 per cent of the respondents said that they had formed community-based organization. Of them 67.8 per cent said that the community based organization has moderately helped in establishing horizontal relationship. It was found that 100 per cent of the respondents said that they had formed community based organization. Of them, 57 per cent said that CBO has moderately helped in establishing vertical relationship. It is to be understood that the sustainability of the project depends on the relationship established by the committee with other organizations and relations that exist within the group, i.e. vertical as well as horizontal. So it can be inferred that CBO has been moderately effective in establishing vertical and horizontal relationships among the participants.

Existence of Committee and its Relationship with Other Organizations

It can be witnessed from the chart that 63 per cent of the respondents were of the view that the committee, apart from associating with the implementing NGO, also has relationship with other bodies or organization, whereas 36 per cent of the respondents said that the committee does not have any relationship with any other organization other than the implementing NGO. It can be witnessed that the NGO

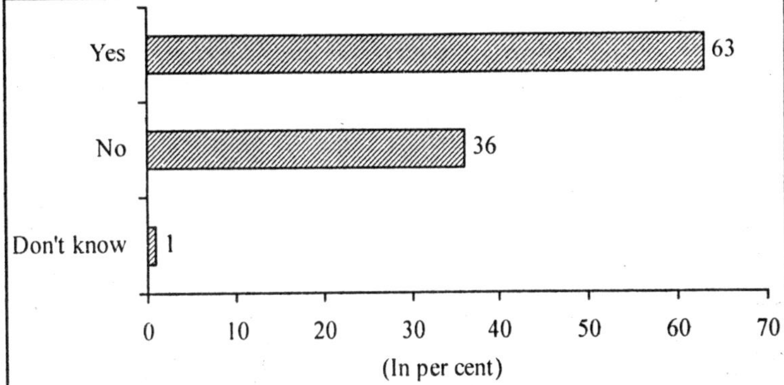

Existence of Relationship with Other Organisation

has been successful in making the committee realize that they need to associate with other organizations to effectively implement the project as it may require different types of expertise.

The analysis revealed that of the total respondents 97 per cent said that, the community-based organization has formed a committee and out of them 63 per cent of the respondents were of the view that the committee had relationship with other organization besides associated NGO. It is to be understood that it is very important for a committee and its members to be associated with other organizations like government and non-governmental agencies so that the resources can be shared and co-ordination could be brought between the activities of such organization/organizations. Such association will help the project and its implementation in the long-run even when the implementing NGO does not extend its association with such community-based organization, i.e. it would make the committee and its members independent and puts them in a position wherein they can see that the committee works on a sustainable-based in the foreseeable future.

6

Analysis and Findings—Part II

This chapter focuses on examining the changes brought in by the agro-forestry project in terms of tangible benefits, viz., material change, social changes, viz., relational change and environmental protection, viz., biological sustainability. In addition, mini cases focusing on the experience and suggestions of beneficiaries have also been put at the appropriate places.

Material Change

Household Income

It can be seen from the adjoining chart that 64 per cent of the respondents are of the view that their household income has increased, whereas 25 per cent of the respondents said that their household

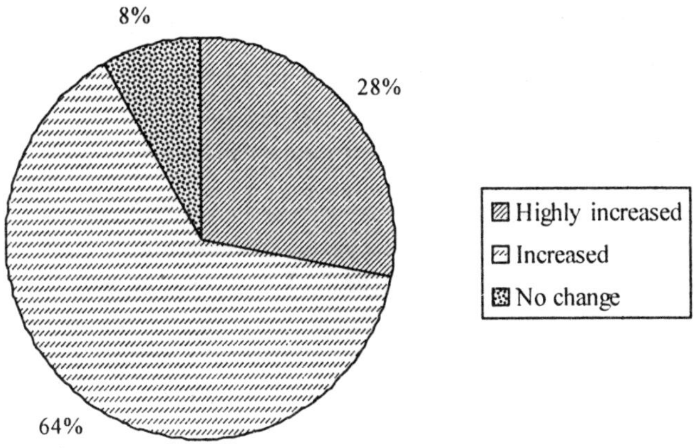

Household Income

income has highly increased. On the other hand, only a small size of respondents said that there has been no change in their household income. It can be perceived that the implementing NGO has not only implemented the project successfully but also made the members realize their expectations from the project, i.e., the project, has increased the household income as said by majority of the respondents.

Decrease in the Migration Levels

It can be seen from the adjoining chart that a majority of the respondents feel that the migration of people to other villages or places has decreased. On the other hand, 43 per cent of the respondents were of the view that the there is no change in the migration of people to other areas. But a minuscule size of the respondents said that the

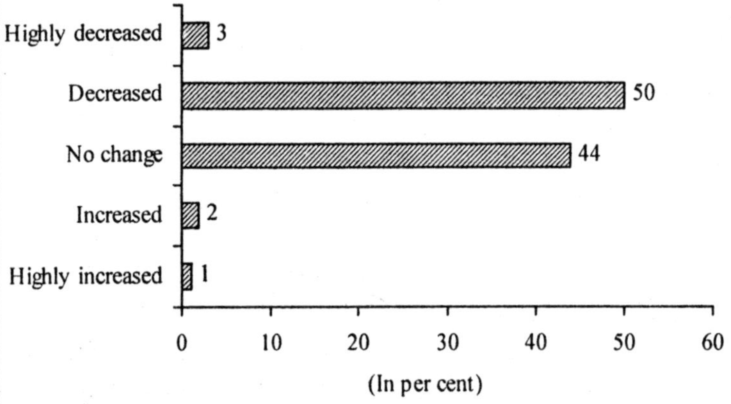

Migration

migration of individuals to other areas had increased. It can be inferred that the project has been successful in reducing that migration of individuals to other areas in search of employment, which reinforces the point that the NGO has been successful in implementing the project and thereby making more number of people get involved.

Mini Case: 1

Beneficiary: Hanumanth Reddy; Village: Chettupally Thanda; NGO: REEDS

Brief Background Information: He is one of the rich farmers among three Reddy families in the village and holds wasteland which is not under cultivation. When the agro-forestry project was implemented in other parts of the village, he decided to take up agro-forestry in his dry and wasteland at the hill area. Since there was no protection for the crop from the animals and birds, he had to stay at the field from morning till evening. Even after erecting some protection, he was not able to protect the crop from wild pigs in the night.

Experience in Plantations: He expressed that this project had been very supportive in all aspects but the only problem which he faced was in the case of supplying water to the crop. But he arranged separate pipes for watering the trees and supplied sufficient water. Of the total 30 trees, now surviving trees were only 17. Of these six were *amla* trees and 10 were *babul* trees.

Opinion on Field and Boundary Plantation: He expressed that this type of cultivation is very useful to all the farmers who own dry lands as this method of cultivation and the trees which are planted require less water when compared to other plants. He cheerfully expressed that this system of agriculture gave him a lot of satisfaction.

He never had hope that there is any scope for the trees to survive in the hilly area. But looking at the survival rate of the trees after six months, he was confident and started concentrating on the crop by providing sufficient manure, fertilizers and water. He told that this system of cultivation in dryland is very useful especially to the common man who is not having water source.

Suggestions as a Practitioner: He suggested that this type of crops in dryland areas is very beneficial for the farmers as most of them have borewells but not with sufficient water in it. It can stop migration as many villagers are migrating to Mumbai. He said he is very keen at doing cultivation and expressed that this sort of project could help change the habits of farmers who are habituated to arrack which takes away a major part of their income.

Agricultural Productivity

It can be seen from the chart that a majority of respondents, i.e. 64 per cent are of the view that the agricultural productivity has increased. On the other hand, 19 per cent of the respondents said that productivity has highly increased. However, a small size of respondents said that there is no change in the agricultural productivity.

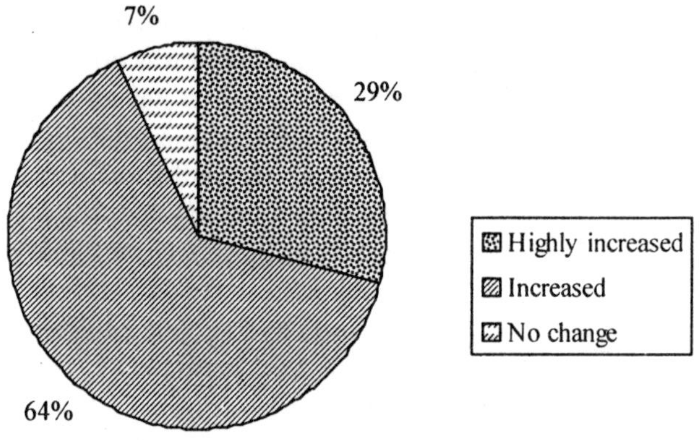

Agricultural Productivity

The analysis reveals the respondents expectations from the project; about 33 per cent of the respondents expected that the project will help them in increasing agricultural productivity. On the other hand, 44 per cent of the respondents took part in the project expecting that it will help them in increasing their household income. It can be seen from the above charts that the project has increased the household income as said by 92 per cent of the respondents. Whereas 92 per cent of the respondents have said that their agricultural productivity has increased or highly increased, from the above analysis it can seen that the implementing NGO has been successful in implementation of the project and meeting the expectations of the members.

It has been observed that about 8 per cent of the respondents expected that the implementation of the project would reduce the migration of individuals to other places. When asked their opinion about the effectiveness of the project in reducing migration 54 per cent of the respondents said that migration has decreased after implementation of the project (from chart named shifting of people).This again reinforces the effectiveness of the project.

Self-development Opportunities

It can be seen from the adjoining chart that 45 per cent of the respondents said that the self-development opportunities have increased. On the other hand, 17 per cent were of the view that the self-development opportunities have highly increased. However, one-third of the respondents said that such opportunities have decreased. On the whole, it can be perceived that the project has increased self-development opportunities as viewed by 64 per cent of the respondents.

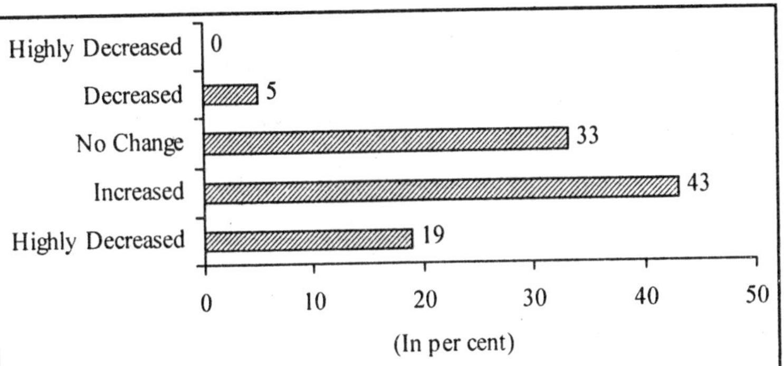

Self-development Opportunities

Village Development

It can be seen from the adjoining chart that 48 per cent of the respondents were of the view that there is an increase in village development. On the other hand, 12 per cent were of the view that there has been a high increase in village development after the implementation of the project. However, 37 per cent of the respondents said that they observed no change in the level of village development. On the whole, it can be said that the project has been successful in developing the village.

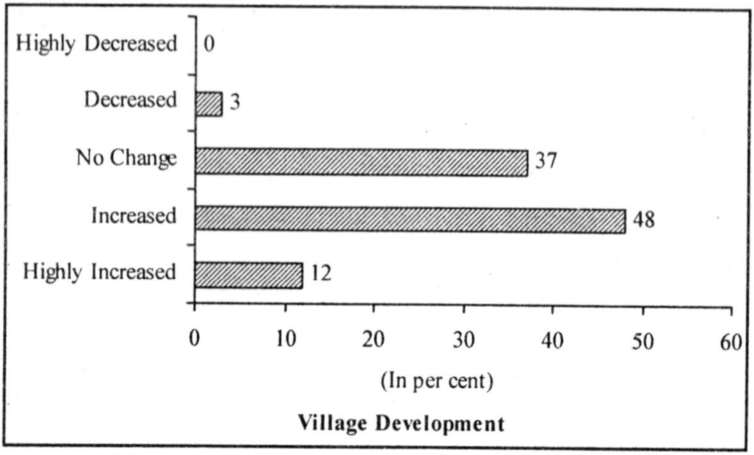

Material Change (Overall)

It can be seen from the adjointing chart that when the social impact of the project is measured in terms of material change on the scale of 1-5, it can be seen that the parameters which help in measuring material change like household income has highly increased (4.2), migration of people has decreased to some extent. On the other hand,

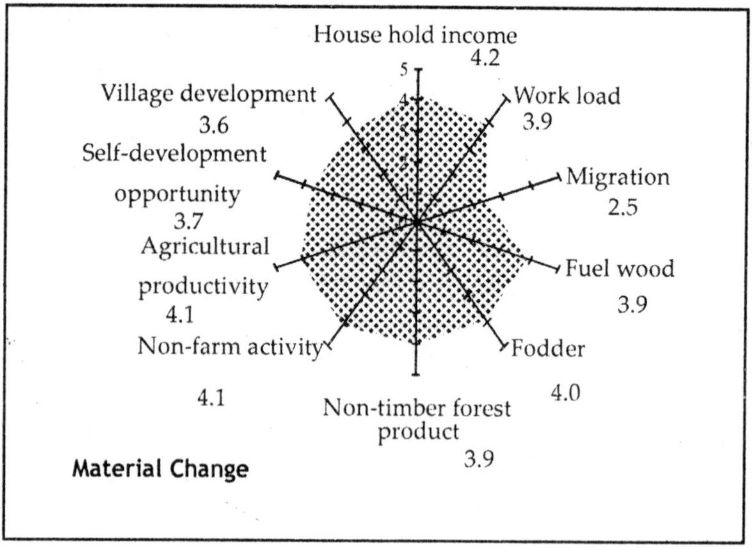

other parameters like village development, self-development opportunities, availability of non-timber forest product, fuelwood has increased. Whereas parameters like agricultural productivity, non-farm activity have highly increased.

Material change is one of the important indicators which reveal the impact of the project on the lives of people which comprise of household income, productivity, fuelwood, fodder availability etc., which has increased after implementation of the project.

It can be seen from the above chart that material change in terms of household income, agricultural productivity, village development and self-development opportunities have highly increased in Nandamalligadda village when measured on the scale of 1-5. Unlike in Nandamalligadda, the material change in above said parameters is on the lower side of the scale. It can be seen from the chart that the

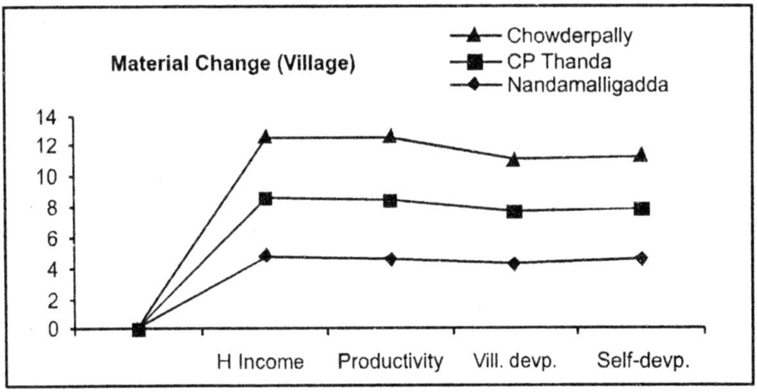

material change with respect to above mentioned parameters has increased or remained unchanged in the other two villages. So it can be stated that the project has been more successful in Nandamalligadda village in case of factors such as household income, agricultural productivity, village development and in providing self-development opportunities. But if we look at overall change in the material factors, the project has been successful in increasing them and there has also been a marked increase in the availability of fuelwood, fodder and non-timber forest products.

Mini Case: 2

Beneficiary: N. Pedda Kistanna Village: Nandimallagadda NGO: SDDPA

Brief Background Information: N. Pedda Kistanna said it was very difficult for him to provide water for the mango crop and other plants before implementation of the project and struggled a lot to see that the trees survive. The reason for high death-survival rate was unperceivable for him even when sufficient water was provided to the crop. He could not find the reason for such low survival rate even after approaching SDDPA staff. Once he had been to Karnataka State on exposure visit that, which was conducted by NGO. It was during this visit he learned about agro-forestry and got interested towards implementing it in his field. He said that practically this method of cultivation is very much result oriented in the long-term and saves money and manpower without investing regularly, when compared to the yearly and seasonal crops. He felt that this type of dry land cultivation made him and others satisfied. Other farmers are also showing interest towards planting these trees in their fields.

Opinion on Field and Boundary Plantation: After implementation of the project he now cultivates vegetables along with mango trees which reduces the temperature in the field thereby increasing the yield when compared to earlier methods of cultivation. He expressed that, earlier to the implementation of the project, he rarely visited the field. But he frequently visits the field as he finds more greenery in the field and he is happy with the survival rate of the boundary plants and teak trees.

Overall Opinion: He expressed that now he is confident of agro-forestry and he now is willing to plant some more plants in the field and enquired from where he could get the plants and what type of species of plants would be useful for his field. He also said that when monsoon starts he would contact forestry department people and get some more plants and if that is not possible he is even ready to buy some plants for his field. He was unhappy with the situation where he himself and other villagers are unaware of the opportunities provided by agro-forestry and sometimes he is discouraged by neighbours as they are also unaware of the benefits which they can reap out of agro-forestry. He told that most of the villagers are also interested in agro-forestry.

Suggestions as a Practitioner: He expressed that as a practitioner this system of cultivation gave him extra income and satisfaction as he can find his field filled with greenery even though he raised that crop in the dryland. Before implementation of the project, most of the farmers were uninterested but after seeing the benefits of the project most of them came forward to implement agro-forestry in their field. But even now the plants which are required are not available.

Mini Case: 3

Beneficiary: N. Ramulu Village: Nandimallagadda NGO: SDDPA

Brief Background Information: Mr. N. Ramulu is a well-established and work-oriented farmer who appreciates the support of NGO and the funding agency as he is satisfied with his mango crop. He purchased a tractor with the funds he has gathered over the last four years by working hard at cultivating the crop in his ten acre land out of which two and a half acres is irrigated land.

Experience in Plantations: He explained that, when he was seedling, he had very less hope, but the NGO gave them good support and they have spent small amount on making pits and in meeting other expenses like trilling, fertilizer, manure and bio-pesticides. The death rate of the plants was more and the water was supplied through the barrels. But even this facility was not used effectively to protect the crop as the water was used for domestic purposes. The reason for high death rate of the plants was unperceivable to everyone.

Opinion on Field and Boundary Plantation: He explained that the mango trees are mostly preferred by other co-farmers. As usual he gave importance to mango crop. He planted three types of trees (mango, lime and tamarind) in three acres of the land of which 75 per cent of the crop is surviving. Out of this crop 95 per cent of the trees are providing good yield and is very much satisfied with the income he has earned from the crop.

Overall Opinion: He expressed that this type of agro-forestry project is very useful for the drylands. He was of the opinion that such projects must be supported by the Government or others as most of the farmers in the village are unaware of agro-forestry, its benefits and even if they are aware they are unable to implement the project due to lack of funds. Most of the farmers ask about agro-forestry and about the plants and they are advised clearly about the project as well as about the plants.

Suggestions as a Practitioner: As a practitioner he felt that this type of cultivation is very beneficial for drylands and this project has provided him with good sum of additional income which he can save for future needs which may not be possible when vegetables are cultivated in the field.

Mini Case: 4

Name: Padma H/o Mylaram Narsimha Village: Chowderpally NGO: PEACE

Brief Background Information: Padma was a wage labour before 1996; her husband used to migrate at least for six months in a year to Bombay. After her father's retirement from the service, he gave her Rs. 20,000. She purchased five acres of land with that money. After some time she came to know that many persons were using that land. Then she complained to the police officer and that problem was not solved. With support from revenue department, she paid them some money and registered that land in her name. The problem was solved with neighbouring farmers. After 1997 she started cultivating on her land.

Experience in Plantations: She participated in the SHG group and took loan from that group. After that she started her live practice in the field. She got a watershed tank near her land for irrigation of her lands and in the summer as tank would get dry the soil from the tank is utilized as a fertilizer for the crop. She has shown more interest towards agro-forestry with implementation of this project. She seeded *Amla,* Tamarind, Cucumber and Teak plants, especially concentrated on custard apple, which were eight in number. Even though with this crop she was gaining some good income yearly for two times, she has undergone training and had been to exposure visits as well. Now she felt happy as tamarind and *amla* trees have started giving yield from current year.

Opinion on Field and Boundary Plantation: She said that the boundary plants, which are around the field, are growing well. These are especially teak trees and custard apple which started giving yield three years back. *Amla* also gave one yield but the crop was very problematic to plants itself because of the more weight of *amla* fruits. So, in earlier stages, she plucked them as the fruits spoiled the crop and even then she got good price for them.

Overall Opinion: She said that many of the farmers are drunkards in their village. They drink everyday and that too local made cheap *arrack,* which affects the nerves and makes them disabled for hard work. Most of the farmers did not show any interest towards understanding the reason for low survival rate of the plants which is the prime reason for decline in the yield of the crop. But now they understand implications of the crop and are interested in seedlings of different crops.

Suggestions as a Practitioner: She said that the agro-forestry is more useful for dryland farmers. This system of agriculture is attracting others to implement agro-forestry as the crop gives good return without much investment of money and manpower.

Relational Change

Social Discrimination

From the adjoining graph more than half, i.e.55 per cent of the respondents accepted that there is no effect by agro-forestry project

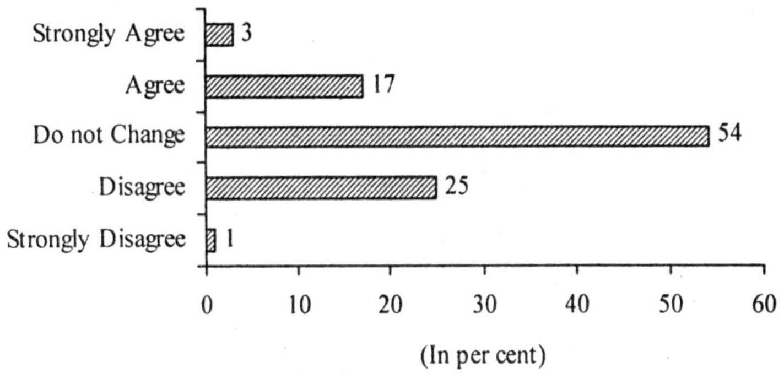

Social Discrimination

on decrease of discrimination on the basis of caste, gender and economical class. About one-fifth, i.e.17 per cent of the respondents are agreed and 3 per cent of respondents are strongly agreed regarding this social discrimination issue that this project has a positive impact to decrease it. More than this one-fourth, i.e. 24 per cent of respondents are disagreed and 1 per cent of the respondent strongly disagree regarding this social discrimination issue that this project has negative impact on decrease of social discrimination.

Participation in Village Development Activities

As seen from the adjoining graph, highest, i.e. 45 per cent respondents said there is no change in participation in village development activities because of agro-forestry project. About enhancement of participation in village development activities majority of the respondents gave negative response, i.e. 27 per cent respondents disagree and 28 per cent strongly disagree that this project is increasing participation in village development activities of people. Not a single respondent said that agro-project is increasing participation in village development activities.

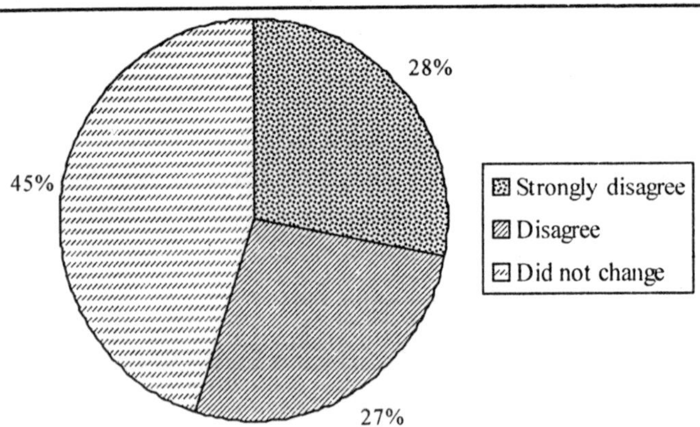

Participation in Village Development Activities

Awareness About the Rights Among People

As can be seen from the adjoining graph, more than half, i.e. 51 per cent of the respondents disagree and 6 per cent of the respondents

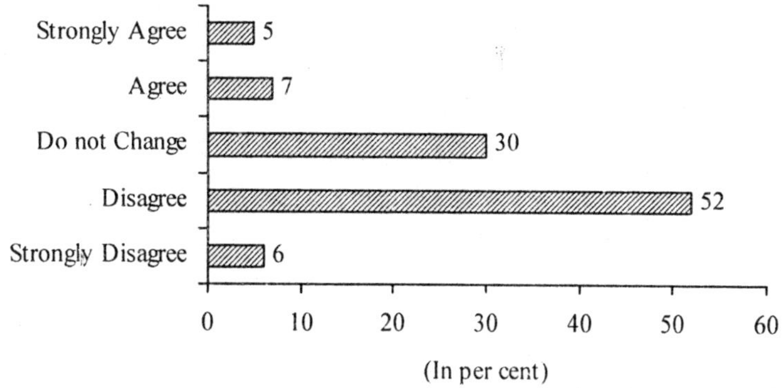

Awareness of Rights

strongly disagree that this project increased awareness of people. We can infer from here that majority of the respondents gave negative response about enhancement of awareness level among people of their rights. Almost one-third of the respondents, i.e. 30 per cent of the respondents said there is no change in awareness level among the people about their rights after implementation of agro-forestry project. Only few a respondents, i.e. 7 per cent of the respondents agreed and 6 per cent respondents strongly agreed that awareness level of their rights increased after implementation of agro-forestry project. We can conclude that a small fraction of the respondents are actually aware of their own rights.

Awareness Level of People about their Responsibility

As can be seen from adjoining graphs majority, i.e. 78 per cent of the respondents disagree and 14 per cent of the respondents strongly disagree that this project increased awareness level of people about their responsibility. We can infer they gave a negative response about enhancement of awareness level of people about their responsibility. Not a single respondent has agreed that this project really had a positive impact on increasing the awareness level of people about their responsibility, whereas a small fraction of population, i.e. 8 per cent believes that this project has not at all affected the awareness level of respondents about their responsibility.

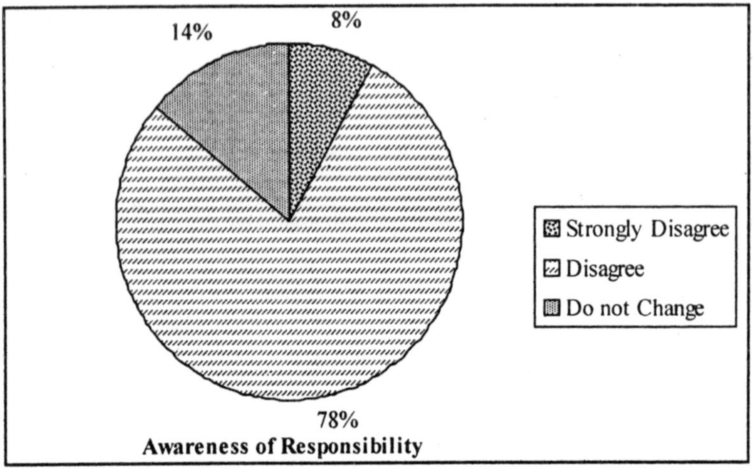

Overall Relational Change

It can be seen from the above graph that the awareness of responsibility among people has not increased and a majority of them have agreed to the same. The same goes with the awareness regarding responsibilities. Here also they were of the view that the project has not been adequately successful in making them realize their responsibility. On the other hand, when it comes to other variables which constitute the part of relational change many agree that there is change in terms of discrimination by caste, awareness of rights,

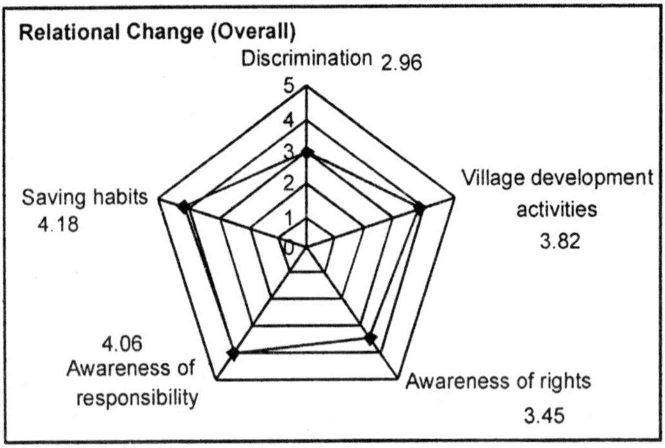

and people's participation in village development activities whereas it can be seen that many respondents were of the view that the project has brought in change in the saving habit of the respondents.

One of the important objective of the project was to bring a change in relational relationship among participants like reducing discrimination by caste, making people participate in village development activities, making people aware of their rights, responsibilities etc. From the above discussion it is clear that the project has not been successful in achieving one of its objectives, i.e. relational changes as there has not been a significant change in the parameters which constitute the part of relational change.

Village-wise Relational Change

It can be seen from the above chart that discrimination by caste, gender and economic class has remained unchanged in all the three villages

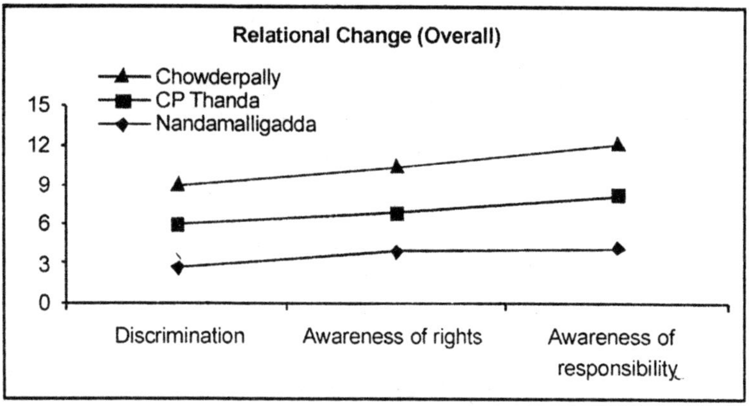

after implementation of the project. When it comes to awareness on rights among the participants, the project has brought a change in Nandamalligadda and Chowderpally villages, but the awareness of rights has remained unchanged in C.P. Thanda village.

Biological Sustainability

Agro-forestry systems bring in distinct benefits and play a vital role in ensuring biological sustainability. Specifically, introduction of agro-forestry helps reduce soil erosion and restores soil fertility of the land. Also, due to the improved vegetative cover, the run-off rainwater is reduced and floods are checked. The dried leaves of trees are used as organic manure that promotes organic farming. The agro-forestry facilitates in percolating down the rainwater in soil, which increases groundwater table. This system helps to maintain and enhance the biodiversity. It gives shelter to plants, animal, birds and soil microbes. Biodiversity is very important for maintaining agricultural productivity. Agro-forestry systems provide fuelwood and fodder. Availability of fuelwood and fodder results in reducing the pressure over forests, which in turns helps restore ecological balance.

Keeping these in mind, the present study tried to examine whether the implemented agro-forestry project was able to bring in the intended benefits of agro-forestry systems from the perspective of biological sustainability as perceived by the respondents. For this, the study included parameters such as species richness, tree density, crown cover, pattern of plantation, availability of water, soil erosion and forest product availability.

The next part of this section tries to analyze the effectiveness of the project in attaining biological sustainability in the villages where the project has been implemented as perceived by the respondents.

Tree Density

As can be seen from the adjoining graph, a majority, i.e. 50 per cent of the respondents and 29 per cent of the respondents said tree density

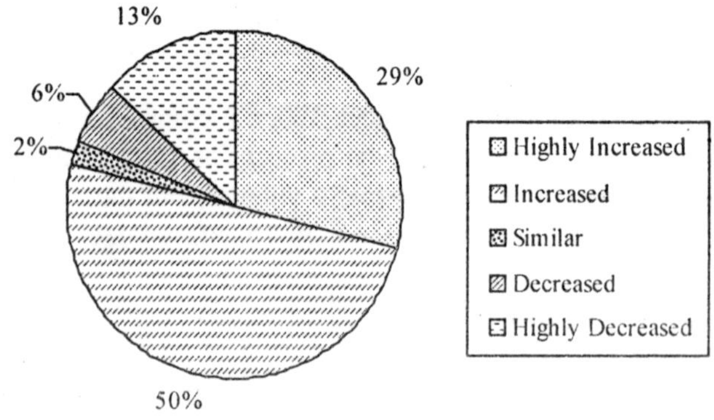

Tree Destiny

has increased and highly increased respectively after implementation of agro-forestry project. Few respondents, i.e. 2 per cent of the respondents said there is a decrease and 6 per cent of the respondents said there is a high decrease in tree density. A small fraction, i.e. 13 per cent of the respondents said tree density has not at all changed because of agro-forestry project. We can conclude that project has a positive affect on tree density as far the opinion of majority of the respondents.

Richness of Species

The adjoining graph shows that a majority, i.e., 67 per cent of the respondents believe that richness of species has increased and 17 per cent of the respondents said it has highly increased after

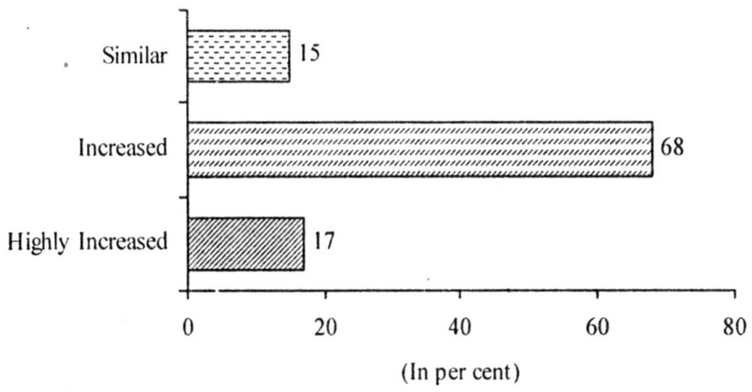

Species Richness

implementation of the project. Not a single respondent was of the opinion that this project has negative impact on richness of species. About one-sixth, i.e. 16 per cent of the respondents said there is no change in richness of species because of this project.

Crown Cover

As can be seen from the adjoining graph, a majority, i.e. 54 per cent of the respondents said crown cover has increased and 8 per cent respondents said crown cover has highly increased. On the other hand,

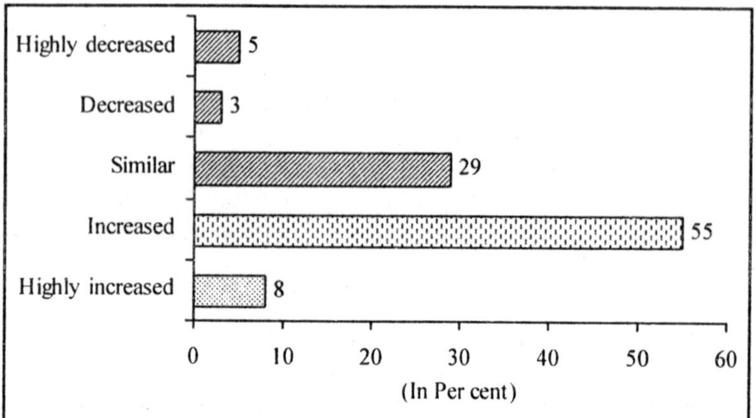

Crown Cover

very small fraction, i.e. 3 per cent of respondents said because of agro-forestry project crown cover have decreased after implementation of agro-forestry project. Almost one-third of the respondents, i.e. 29 per cent of the respondents have said there is no change in crown cover because of this project. We can conclude that a majority of the respondents are of the opinion that this project has positive impact on crown over and respondents who said project has not at all affected crown cover is of a significant number.

Overall Change in Biological Sustainability

It can be seen from the chart that when the social impact of the project is measured in terms of biological sustainability on the scale of 1-5, it can be seen that the parameters which help in measuring biological sustainability like plantation pattern, tree density, species richness have increased and parameters like forest product availability, crown cover and water availability have remained unchanged. Biological sustainability is one of the important indicators which reveals the impact of the project on biological sustainability which comprise forest product availability, species richness, tree density, crown cover, plantation pattern, availability of water and soil erosion.

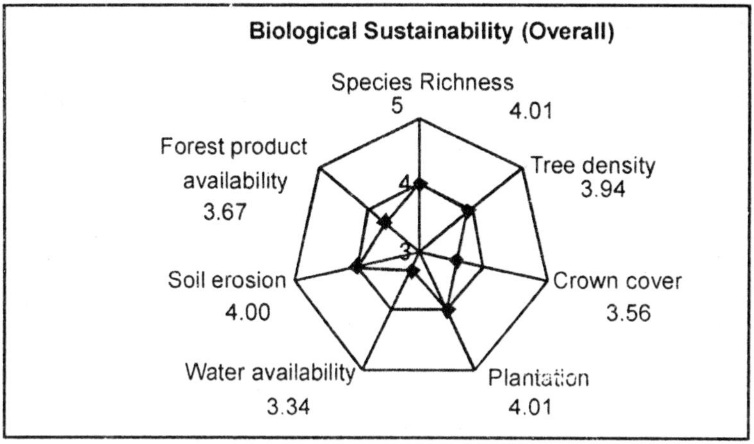

Village-wise Biological Sustainability

It can be seen from the above chart that biological sustainability in terms of species richness has increased in all the three villages after implementation of the project. When it comes to biological sustainability in terms of water availability, it has increased in

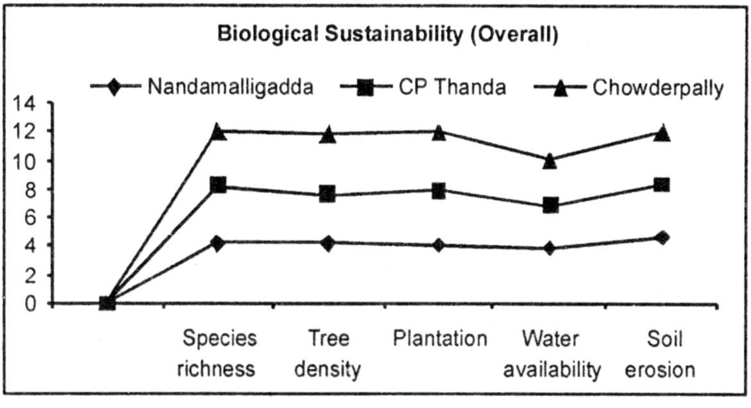

Nandamalligadda and remained unchanged in C.P. Thanda and Chowderpally villages, whereas in case of soil erosion it has highly increased in Nandamalligadda village and has increased in C.P. Thanda and Chowderpally villages.

Mini Case: 5

Name: Manikya Nayak Village: Bolli Bai Tanda NGO: REEDS

Brief Background Information: The farmer belongs to a scheduled tribe and nuclear family, having twenty acres of land of which ten acres are irrigated and ten acres are rainfed. The land was in a dry condition and as he did not have sufficient income in those days he decided to plant different kinds of trees through the project and showed interest in the growth of those trees. Now he feels very happy when we ask him about the agro-forestry (his trees) and he tells that the crop is giving him supplementary income and he is also earning profits for two years. He felt very happy that the utilization of the dry land in two crops at a time is giving him enormous satisfaction and he feels that there is some greenery to visit his field throughout the year.

Experience in Plantations: He planted trees on four acres in total, of which 45 are Mango trees, 20 Tamarind and 30 Subabul trees. Now he is having 22 Mango, 12 Tamarind and 20 Subabul trees. He told that remaining trees are dead only because of lack of water due to less supply of electric power supply and due to scarcity of the water. The electric supply was in fluctuations and was dim most of the time, which made it impossible for him to provide sufficient water to the crop. He told that he talked about this problem with other farmers but they were never interested in solving it. He explained that the CBO has not shown interest and, even when some farmers were provided water in barrels, the water was used for domestic purpose. This was the major cause for lack of sufficient water supply. The problem was mainly with animals and goats, since it was very difficult to protect the crop from them as the plants were very small in size and the investment from their side and from the NGO was very small.

Opinion on Field and Boundary Plantation: He told that the field was not in the use earlier but, after the trees were planted, there is some fodder availability for the animals and some small crop is also being cultivated and the boundary trees are also more useful in protecting the crop from the animals. He felt that he had committed a mistake by not planting teak trees as they were grown in large numbers in other lands in those days.

Overall Opinion: He expressed that now he is interested in planting tress on other four acres and other farmers as well expressed their willingness towards planting mango trees in their fields as it can bring the dealers to the village itself and then it is not required for them to go to the market to sell their produce. He explained that most of the farmers hold drylands in which they are not doing any cultivation. He wanted the NGO to support all such farmers as all of them are interested in implementing agro-forestry in their field.

Mini Case: 6

Beneficiary: Laxmikanth Reddy *Village: Chilumulamularam*
NGO: REEDS

Brief Background Information: Laxmikanth Reddy holds large area of land in the village which is not under cultivation. He decided to implement agro-forestry in his land and witnessed a substantial improvement in the produce and life of the plants (50 Mango trees are alive). This area now has sufficient irrigation facilities and survival of the trees has substantially improved thereby reducing the expenditure on the crop. He has another 45 acres of land which is rainfed and 3 acres of land with irrigation facilities.

Experience in Plantations: He implemented agro-forestry in four and half acres, total trees were 300 but out of these trees only 50 are surviving because of the soil fertility. Though he could not protect all those trees in the last two years, he expressed that the crop has given sufficient yield and earned him ten thousand rupees in the last two years. After implementing the agro-forestry project, the field has total boundary.

Opinion on Field and Boundary Plantation: He expressed that the loss of crop can be attributed to absence of boundary and also due to low fertility of the soil. The major part of the lost crop was the crop present on the boundary. He expressed that this system of agriculture is very useful for the small agriculturists in the absence of proper irrigation facilities. He was not confident of the success of the agro-forestry project before implementation. But out of experience he has learnt that it is really beneficial and helps in increasing the yield of the crop. He now suggests other villagers to implement agro-forestry in their fields as well to enjoy the benefits which he has reaped from the project.

Overall Opinion: He was of the opinion that many more people are interested in agro-forestry than ever before and a majority has realized that the project can reap supplementary income apart from keeping the crop growing throughout the year. He also said that he would plant more trees in the near future if any such project is implemented with someone's support and he is willing to bring in more than 7 acres into cultivation if such project is taken up in their village. He expressed that with small investments farmers can earn good profit apart from enjoying other benefits.

Suggestions as a Practitioner: He advised the farmers to implement this sort of projects especially in drylands which lack soil fertility, irrigation facilities and electric power supply. Small farmers in his village are migrating to other places in search of livelihood due do low yield of their crop. He expressed that the farmers will be benefited if such projects are continued in the near future.

Mini Case: 7

Beneficiary: Bhimla Nayak Village: Chettupally Tanda NGO: REEDS

Brief Background Information: Mr. Bhimal Nayak belongs to scheduled tribe and lives in a joint family. He owns 37 acres of land of which 7 acres has irrigation facility and the remaining land is rainfed.

Experience in Plantations: He implemented agro-forestry in four and half acres, total trees were 300 but out of these trees only 50 are surviving because of the soil fertility. Though he could not protect all those trees in the last two years, he expressed that the crop has given sufficient yield and earned him ten thousand rupees in the last two years trees. After implementing the agro-forestry project the field has total boundary.

Overall Opinion: He expressed that the loss of crop can be attributed to absence of boundary and also due to low fertility of the soil. The major part of the last crop was the crop present on the boundary. He expressed that this system of agriculture is very useful for the small agriculturists in the absence of proper irrigation facilities. He was not confident of the success of the agro-forestry project before implementation. But, out of experience, he has learnt that it is really beneficial and helps in increasing the yield of the crop. He now advises other villagers to implement agro-forestry in their fields as well to enjoy the benefits which he has reaped from the project.

Suggestions as a Practitioner: He suggested the farmers to implement this sort of projects especially in drylands which lack soil fertility, irrigation facilities and electric power supply. Small farmers in his village are migrating to other places in search of livelihood due do low yield of their crop. He expressed that the farmers will be benefited if such projects are continued in the near future.

7

Analysis and Findings—Part III

This chapter focuses on examining the perception and feedback for promotion of agro-forestry system, training programmes, capacity and skill development, knowledge improvement, performance of NGO, extent of expectation met, empowerment, effectiveness of project and approach adopted. For an indepth analysis, cross-tabulation, hypothesis testing and Analysis of variance have also been performed to gain added insights. In addition, mini cases focusing on the experience and suggestions of beneficiaries have also been put at the appropriate places.

Perception and Feedback

Willingness to Promote Agro-forestry System

It can be seen from the chart that, of the total respondents, 52, i.e. a majority of the respondents were willing to promote agro-forestry

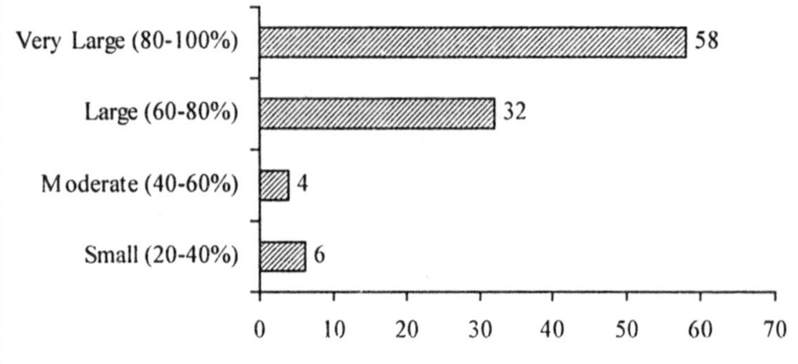

Willing to Promote Agro-forestry

project to a very large extent. On the other hand, 32 per cent of the total respondents said that they will promote the agro-forestry project to large extent, whereas a minuscule part of total respondents, i.e.10 per cent of the total respondents said that they will promote the agro-forestry project to a moderate or small extent. It can be perceived that a majority of respondents are readily willing to promote agro-forestry project which would motivate more number of people implement to this practice and increase the chances of the project's sustainability in the long-run.

Opportunity to Participate in Trainings

As can be seen in the adjoining graph, 47 per cent of the respondents responded positively and were convinced that most of the villagers were provided with the opportunities to participate in trainings organized by NGO, whereas 21 per cent of the respondents said that a majority of the villagers had opportunities to participate in trainings organized by NGO, 8 per cent of the respondents were moderately convinced in saying that the villagers had opportunities to participate in trainings organized by NGO, 12 per cent of the respondents were least convinced in saying that the villagers had opportunities to participate in trainings organized by NGO, 12 per cent respondents felt that very few villagers had opportunities to participate in trainings organized by NGO. It goes on to say that many of the respondents responded positively, but we cannot ignore the other half of the respondents too. Training of the members with respect to skill

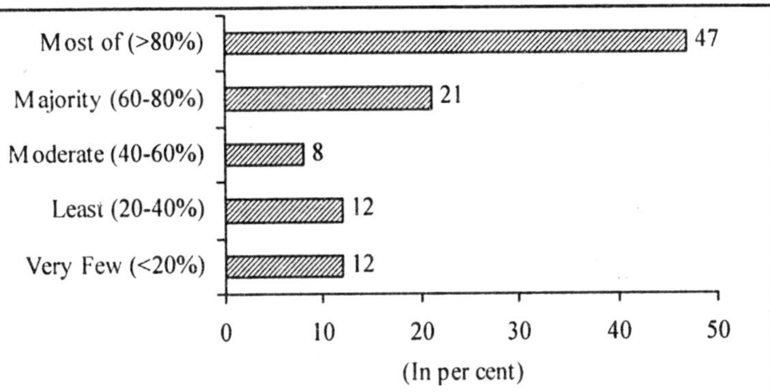

Opportunity to Participate in Training Session

development will not only help in increasing the sustainability of the project but also helps in increasing the efficiency and effectiveness with which one works towards increasing the productivity or towards increasing the household income levels. It needs to be specified that training is one of the means of replacing the NGO in the long-run and making the participants capable of continuing the project which makes the project sustain in the long-run.

Village-wise Opportunity to Participate in Training Programmes

It can be seen from the above chart out of the respondents who belong to Nandamalligadda village, 90 per cent said that they had the opportunity to participate in training programmes most of the times and the remaining 10 per cent of them said that they had the opportunity to participate such programmes majority of times. On the other hand of the respondents from C.P. Thanda village 36.6 per cent said that they had very few opportunities to participate in training programmes, 33.30 per cent said they had least chances to participate in such programmes and just about 20 per cent of them

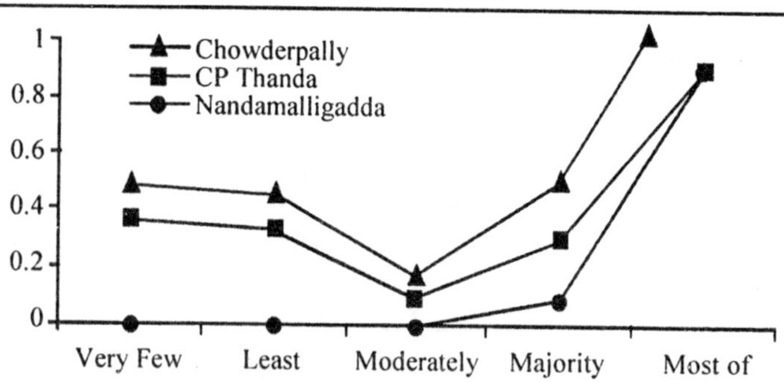

Opportunity to Participate in Training Programmes (Village-wise)

were of the view that majority of times they had the opportunity to participate in such programmes whereas of the respondents from Chowderpally village 12.20 per cent of the respondents said that they had very few or least chances to participate in training programmes respectively, 21.10 per cent of the respondents said that majority of the times they had the opportunity to participate in such programmes and about 46 per cent of the respondents said that they had the opportunity to participate in village development activities most of the times.

To assess the efficacy of the project with respect to providing equal opportunity for the respondents of all the three villages to participate in training sessions, analysis of variance was done. The hypotheses proposed for the analysis were:

Null Hypothesis—H_0: There is no significant difference in the opportunities provided to the respondents to participate in training programmes in all the three villages.

Alternative Hypothesis—H_1: There is a significant difference in the opportunities provided to the respondents to participate in training programmes in all the three villages.

The **probability-value** for this test is **0.0000**. Also the observed *F*-value falls in the rejection zone, at a significance level of 5 per cent. Hence we reject the null hypothesis. Alternatively, we accept that there is a significant difference between the opportunities provided to participants of all the three villages. This implies that, as per the perception of the respondents there is a significant difference in the opportunity provided to them to participate in training programmes. Alternately, it can be inferred that *the respondents were not provided equal opportunity to participate in the training pragrammes.*

Effectiveness of Training Programmes

As can be seen in the adjoining graph, majority of respondents, i.e., 65 per cent responded positively and were convinced that the training

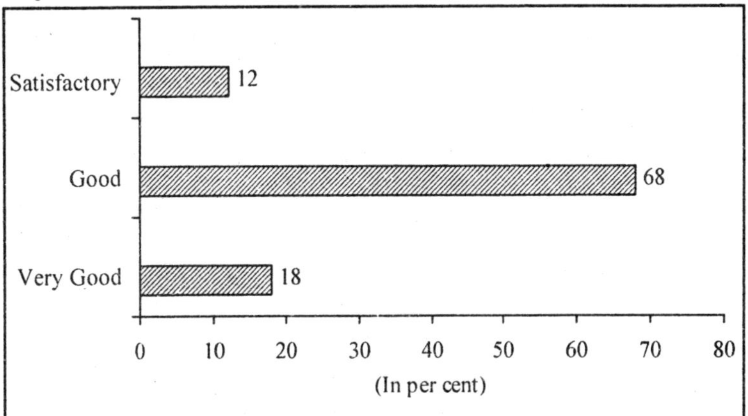

Feasibility of Training Programme

programmes are very useful for capacity development, whereas 27 per cent of the respondents, responded saying that the training programmes are useful for capacity development, 7 per cent of the respondents found no change in capacity development as a result of the training programmes, 1 per cent of the respondents found training programmes to be useless. It goes on to say that many of the respondents responded positively and found the training programmes of some use in capacity development.

Village-wise Capacity Development

It can be seen from the adjoining chart that, of the total respondents from Nandamalligadda village 80 per cent said that, the training programmes are very useful for capacity development. On the other hand, of the total respondents from C.P. Thanda village, 53.3 per cent of the respondents were of the view that, the training programmes are very useful and 30 per cent said that such programmes are useful for capacity development. Where as of the total respondents 65.60 per cent of the respondents said that such programmes are very useful for capacity development.

Capacity Development (Village-wise)

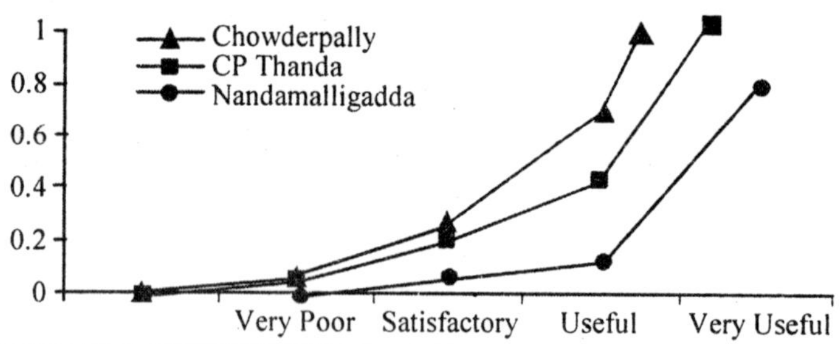

Opportunity of Participation in Training Programmes and its Impact

It can be seen from the table that of the total respondents who said that most of the times they had the opportunity to participate. 73.8 per cent said that the training programmes are very useful for capacity development and the remaining 21.4 per cent and 4.8 per cent of the

Opportunities Provided to Participate in Training Programme	*Usefulness of Training Programme for Capacity Development*			
	Very poor	*Satisfactory*	*Useful*	*Very Useful*
Very Few			27.3%	72.7%
Least	9.1%	18.2%	9.1%	63.6%
Moderately			57.1%	42.9%
Majority		10.5%	36.8%	52.6%
Most of	1.1%	6.7%	26.7%	65.6%

respondents of this category were of the view that the training programmes are useful and satisfactory for capacity development. On the other hand of the respondents who said that majority of the times they had got an opportunity to participate in training programmes. 52.6 per cent said that the training programmes are very useful and about 36.8 per cent of this category said that the programmes are useful for capacity development whereas 9.1 per cent of the respondents who said that they have got least opportunity to participate in the training programmes. 63.6 per cent said that these programmes are very useful for capacity development and of the total respondents who were of the view that they have got very few opportunities to participate in training programmes. A majority said that such programmes are very useful for capacity development. It can be perceived that about 68 per cent of the total respondents said that they have got a fair opportunity to participate in such programmes and of the total respondents 92 per cent said that such programmes are useful for capacity development.

It can be understood that the project or the implementing NGO was not successful at providing fair opportunity for all the respondents even though 68 per cent of the respondents said that they have got fair opportunity to participate in such programmes. When it comes to the usefulness of these programmes for capacity development it cannot be denied that NGO has been successful in making the training programmes effective. But to end it needs to be borne in mind that a fair chance must be provided to all the participants to make it more effective and sustainable in the long-run.

Skill Development

As can be seen in the adjoining graph, majority of respondents, i.e., 68 per cent, responded positively and responded saying that the skill

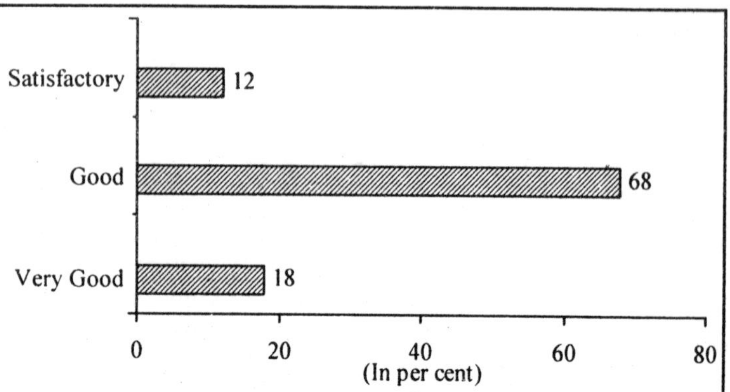

Skill Development

		Skill Development		
		Satisfactory	*Good*	*Very good*
Village	Nandamalligadda	3.3%	53.3%	43.3%
	CP Thanda	16.7%	83.3%	0%
	Chowderpally	16.7	66.7%	16.7%

development in him was very good, whereas 20 per cent of the respondents responded saying that there was good skill development in him, 12 per cent of the respondents responded saying that they had satisfactory skill development in them. But none of the respondents said that they had poor or very poor skill development in them. It goes on to say that, the majority of the respondents believed that they had certain level of skill development in them.

It can be seen from the adjacent table that of the total respondents of Nandamalligadda 43.3 per cent said that there has been very good development in the skill of the respondents after implementation of the project and 53.3 per cent feel that the skill development has been good. On the other hand 83.3 per cent of the respondents of C.P. Thanda said that the skill development of the respondents has been good after implementation of the project whereas 66.7 per cent of the respondents of Chowderpally were of the view that the skill development has been good after implementation of the project.

To assess the efficacy of the project with respect to skill development of the respondents of all the three villages, Analysis of Variance was done. The hypothesis proposed was:

Null Hypothesis—H_0: There is no significant difference in the level of skill development in the respondents of all the three villages.

Alternative Hypothesis—H_1: There is a significant difference in the level of skill development of the respondents of all the three villages.

The **probability-value** for this test is **0.0000**. Also the observed F-value falls in the rejection zone, at a significance level of 5 per cent. Hence we reject the null hypothesis. Alternatively, we accept that there is a significant difference in the skill development of participants of all the three villages. This implies that, as per the perception of the respondents there is a significant difference in the level of skill development of the respondents. Alternatively, it can be inferred that *the level of skill development of the participants differed from village to village.*

To assess whether there exists any dependency between educational level and skill development of the respondents, Chi Square test was done and the hypothesis proposed were:

Null Hypothesis—H_0: The skill development and education level are independent.

Alternative Hypothesis—H_1: Skill development depends on education level of the respondents.

The **probability-value** for this test is **0.70**. Also the observed value falls in the acceptance zone, at a significance level of 5 per cent. Hence we accept the null hypothesis. Alternatively, we accept that the skill development and educational level are independent of each other. This implies that, as per the perception of the respondents the skill development and education level are independent. Alternatively, it can be inferred that *the level of skill development of the participants is independent of the educational level of the respondents.*

Extent of Knowledge Improvement in Villagers

As can be seen in the adjoining graph, 30 per cent of the respondents responded saying that the extent of knowledge improvement in villagers was very good, whereas 36 per cent of the respondents responded saying that there was good knowledge improvement in villagers, 34 per cent of the respondents responded saying that extent of knowledge improvement in villagers was satisfactory. But none of the respondents said that there was poor or very poor knowledge improvement in villagers. It goes on to say that, the majority of the

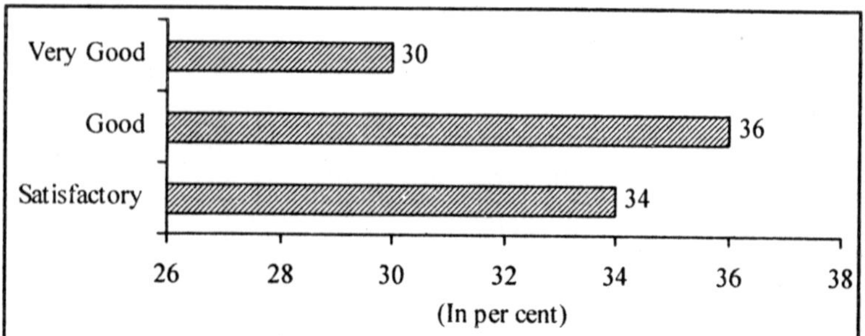

Knowledge Development (Overall)

respondents believed that there is certain level of knowledge improvement in villagers.

Village-wise Knowledge Improvement

It can be seen from the adjacent graph that of the total respondents of Nandamalligadda 60% said that there has been very good improvement in the knowledge of the respondents after implementation of the project and 33.33% feel that the knowledge improvement has been good.

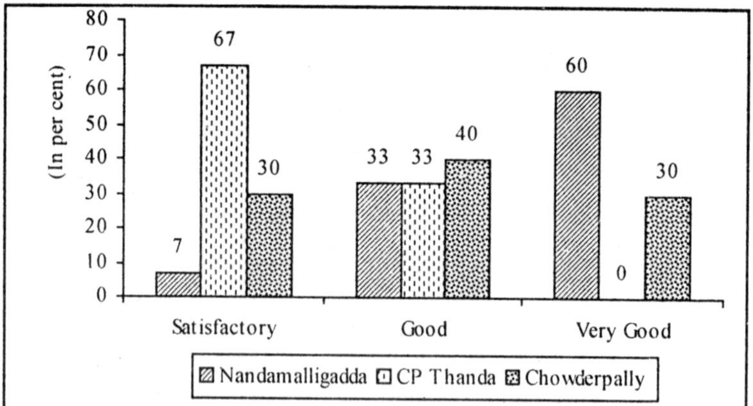

Knowledge Development (Village-wise)

On the other hand 66.7 per cent of the respondents of C.P.Thanda said that the knowledge improvement of the respondents has been satisfactory after implementation of the project whereas 40 per cent of the respondents of Chowderpally were of the view that the knowledge improvement has been good after implementation of the project.

To assess the efficacy of the project with respect to knowledge improvement of the respondents of all the three villages, Analysis of Variance was done and the hypotheses proposed were:

Null Hypothesis—H_0: There is no significant difference in the level of knowledge improvement of the respondents of all the three villages.

Alternative Hypothesis—H_1: There is a significant difference in the level of knowledge improvement of the respondents of all the three villages.

The **probability-value** for this test is **0.0000**. Also the observed *F*-value falls in the rejection zone, at a significance level of 5 per cent. Hence we reject the null hypothesis. Alternatively, we accept that there is a significant difference in the knowledge improvement of participants of all the three villages. This implies that, as per the perception of the respondents there is a significant difference in the level of knowledge improvement of the respondents. Alternatively, it can be inferred that *the level of knowledge improvement of the participants differed from village to village.*

To assess whether there exists any dependency between educational level and knowledge improvement of the respondents, Chi Square test was done and the hypothesis proposed were:

Null Hypothesis—H_0: The knowledge improvement and educational level are independent.

Alternative Hypothesis—H_1: Knowledge improvement depends on educational level.

The **probability-value** for this test is **0.30**. Also the observed value falls in the acceptance zone, at a significance level of 5 per cent. Hence we accept the null hypothesis. Alternatively, we accept that the knowledge improvement and educational level are independent of each other. This implies that knowledge improvement and educational level are independent. Alternatively, it can be inferred that *the level of knowledge improvement of the participants is independent of the educational level of the respondents.*

Performance of NGO

Rating of NGO's Performance

As can be seen in the adjoining graph, majority of respondents, i.e., 58 per cent, responded saying that they would rate performance of

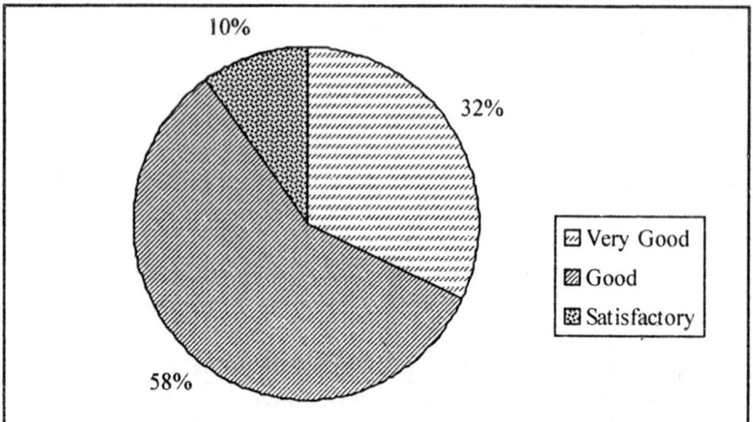

NGO Performance (Overall)

NGO as good, whereas 32 per cent of the respondents responded saying that the performance of the NGO is very good, 10 per cent of the respondents found it to be satisfactory. But none of the respondents said that the performance of the NGO is bad or very bad. It goes on to say that, the majority of the respondents believed that there is certain level of knowledge improvement in villagers.

Village-wise NGO Performance

From the adjoining chart, it can be discerned that when on a five point rating scale, where five corresponded to very good and one corresponded to very bad, the responses from the beneficiaries were solicited regarding the performance of implementing NGO,

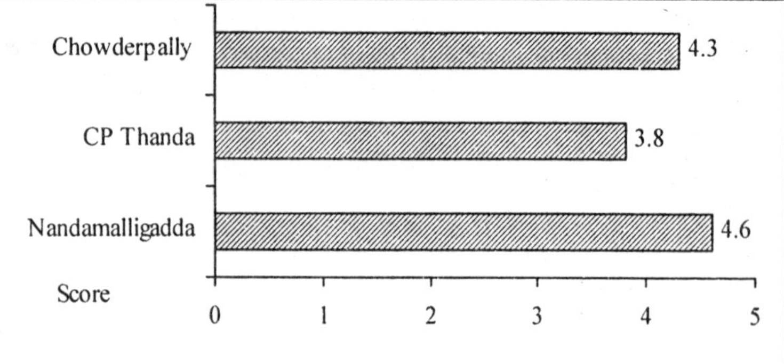

Village-wise Performance of NGO

Nandamalligadda emerged as the best among the three. So, it can be said that the NGO implementing the agro-forestry project there have been very effective in doing so from the perspective of beneficiary farmers. At the same time, NGO implementing the project at C.P. Thanda got the least score among the three suggesting that the farmers of C.P. Thanda did not view the work done by them as very effective.

To assess the efficacy of the project with respect to performance of the implementing NGO in all the three villages, Analysis of Variance was done and the hypotheses proposed were:

Null Hypothesis—H_0: There is no significant difference in the performance of the implementing NGO in all the three villages.

Alternative Hypothesis—H_1: There is a significant difference in the performance of the implementing NGO in all the three villages.

The **probability-value** for this test is **0.0000**. Also the observed *F*-value falls in the rejection zone, at a significance level of 5%. Hence we reject the null hypothesis. Alternatively, we accept that there is a significant difference in the performance on the implementing NGO in all the three villages. This implies that, as per the perception of the respondents there is a significant difference in the performance of implementing NGO in all the three villages. Alternatively, it can be inferred that *NGO's performance differed from village to village.*

Mini Case: 8

Beneficiary: Sani Yadagiri Village: Chowderpally NGO: PEACE

Brief Background Information: He holds four acres of land of which two acres is irrigated and other two acres is rainfed. Besides the cultivation he does tailoring work in the village and his income is satisfactory. With the help of this agro-forestry, he now has more benefits because he planted more plants and satisfactorily told that these crops are well and even boundary trees are well around the field.

Experience in Plantations: He said that he was not interested in these crops and with NGO advice he started seedling of the plants. After that even other farmers showed interest and family members also gave them importance. After one year he felt confident that the crop will do better and started showing keen interest. After implementation of the project even the family members supported and helped him in watering and manuring the crop. Now they are also satisfied. The supplying water in the earlier stages was difficult because there was no rain and because monsoon itself was not there.

Opinion on Field and Boundary Plantation: He planted 85 mango trees in two acres, 90 custard apple plants, 230 teak plants and 100 jetropa plants. He said that this system of cultivation is more useful in dry land areas. Now the teak trees which are in the boundary are healthy and most of them are grown with strong stems.

Overall Opinion: He said that in case of failure of monsoon these crops would give some income to farmers for survival and not require any protection from animals like other crops. He expressed these plants are in natural system and gives yearly crops. Yadagiri told that other villagers are now willing to plant these sorts of trees in their fields.

Suggestions as a Practitioner: As a practitioner he views that plantation is very easy as compared to growing crops. These plants do not require much expenditure and manpower. He said if the villagers plant more of mango trees, brokers would come to village to buy from the village. Thus they can reduce their transaction cost also; there will be no need for them to go to cities for selling their produce.

Empowerment

Social Empowerment

As can be seen in the adjoining graph, only 19 per cent of the respondents responded saying that there has been very high social empowerment after the initiation of Agro-forestry project, whereas 45 per cent of the respondents responded saying that there was high social empowerment after the initiation of Agro-forestry project, 36 per cent of the respondents responded saying that there is no change in social empowerment after the initiation of Agro-forestry project. But none of the respondents said that there is poor or very poor Social empowerment after the initiation of Agro-forestry project. It goes on to say that, the majority of the respondents believed that there is certain level of social empowerment after the project.

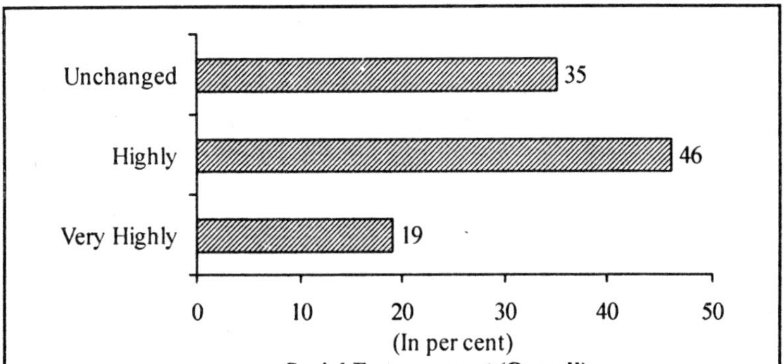

Social Empowerment (Overall)

Village-wise Social Empowerment

It can be seen from the adjoining chart that 76.6 per cent of the respondents of C.P. Thanda village were of the view that there has been no change in social empowerment of the people, 20 per cent feel that social empowerment has highly increased. On the other hand the picture seems to complete opposite when it comes to Chowderpally village as 73.3 per cent of the respondents from this village feel that social empowerment has highly increased and just about 26.7 per cent feel that there has been no change in social empowerment whereas 53.3 per cent of the respondents of Nandamalligadda village were of the view that social empowerment has highly increased and 43.3 per cent said that social empowerment has highly increased.

To assess the efficacy of the project with respect to social empowerment in all the three villages, Analysis of Variance was done and the hypothesis proposed were:

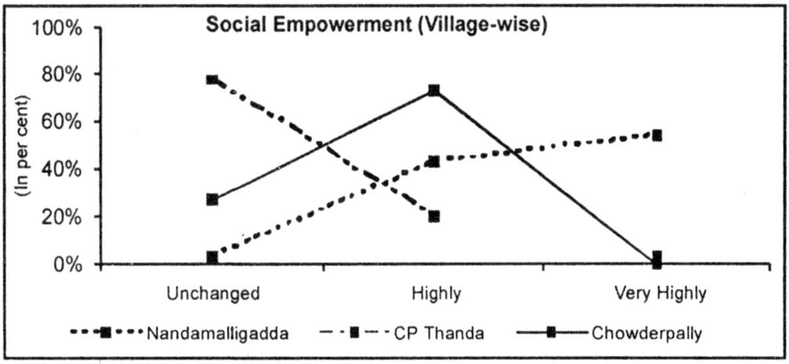

Null Hypothesis—H_0: There is no significant difference in social empowerment of the respondents of all the three villages.

Alternative Hypothesis—H_1: There is a significant difference in social empowerment of the respondents of all the three villages.

The **probability-value** for this test is **0.0000**. Also the observed F-value falls in the rejection zone at a significance level of 5 per cent. Hence, we reject the null hypothesis. Alternatively, we accept that there is a significant difference in the social empowerment of the respondents of all the three villages. This implies that, as per the perception of the respondents there is a significant difference in the social empowerment of the respondents of all the three villages. Alternatively, it can be inferred that *Social Empowerment of the respondents differ from village to village.*

Economic Empowerment

It can be observed from the above graph that only 14 per cent of the respondents responded saying that there has been very high economic empowerment after the initiation of Agro-forestry project, whereas 55 per cent of the respondents responded saying that there was high economic empowerment after the initiation of Agro-forestry project, 29 per cent of the respondents responded saying that there is no change in economic empowerment after the initiation of Agro-forestry project, 2 per cent of the respondents found poor economic empowerment after the initiation of Agro-forestry project. But none of the respondents said that there is very poor economic empowerment after

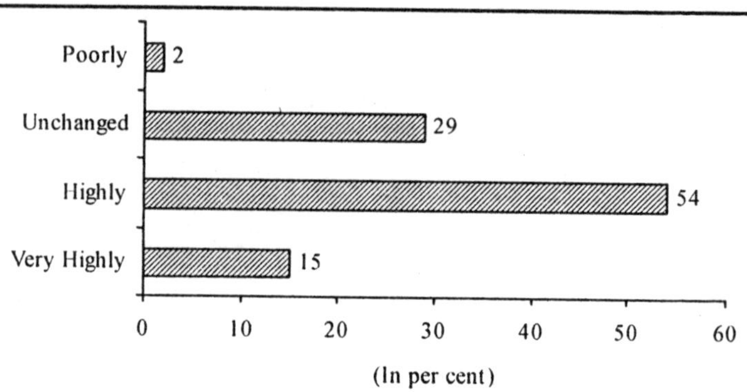

Economic Empowerment (Overall)

the initiation of Agro-forestry project. It goes on to say that the majority of the respondents believed that there is certain level of economic empowerment after the initiation of Agro-forestry project.

The adjoining graph depicts the positive effect of economic empowerment in the three villages. Among the three villages respondents of Nandamalligadda had given positive response. Not even a single respondent gave a negative response about economic empowerment in Nandamalligadda. C.P. Thanda is the only village

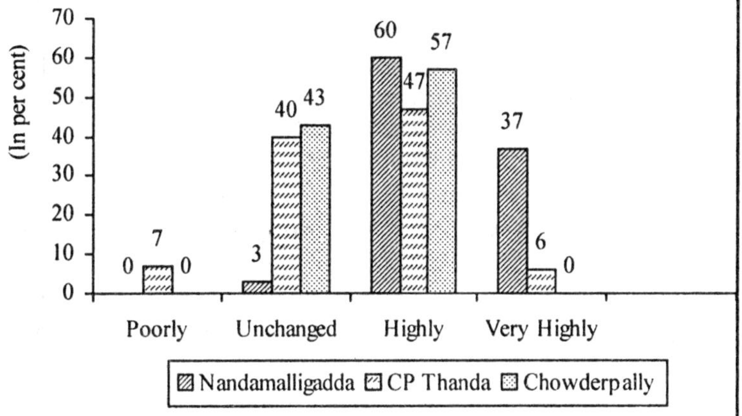

Economic Empowerment (Village-wise)

where few respondents, i.e. 6.66 per cent gave negative response. We can infer that in all the villages, majority of the respondents perceived that economic empowerment was affected positively after commencement of project.

To assess the efficacy of the project with respect to economic empowerment in all the three villages, Analysis of Variance was done and the hypotheses proposed were:

Null Hypothesis—H_0: There is no significant difference in economic empowerment of the respondents of all the three villages.

Alternative Hypothesis—H_1: There is a significant difference in economic empowerment of the respondents of all the three villages.

The **probability-value** for this test is **0.0000**. Also the observed F-value falls in the rejection zone, at a significance level of 5 per cent. Hence we reject the null hypothesis. Alternatively, we accept that there is a significant difference in the economic empowerment of the respondents of all the three villages. This implies that, as per the perception of the respondents there is a significant difference in the economic empowerment of the respondents of all the three villages. Alternatively, it can be inferred that *economic empowerment of the respondents differ from village to village.*

Overall Empowerment of Villagers

As can be observed from the graph below, only 18 per cent of the respondents responded saying there was very high overall empowerment after the initiation of Agro-forestry project, whereas 52 per cent of the respondents responded saying there was high overall empowerment after the initiation of Agro-forestry project, 29 per cent of the respondents responded saying there was no change in overall empowerment after the initiation of Agro forestry project, 1 per cent of the respondents found poor overall empowerment after the initiation of Agro-forestry project. But none of the respondents said that there is very poor overall empowerment after the initiation of Agro-forestry project. It goes on to say that, the majority of the respondents believed that there is certain level of overall empowerment after the initiation of Agro forestry project.

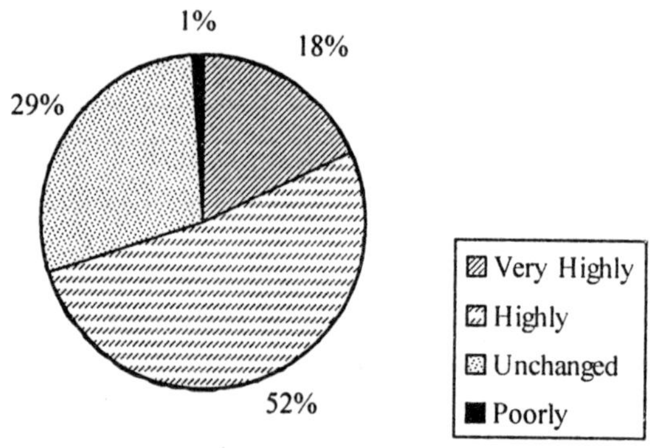

Overall Empowerment

It is clear from the adjoining graph that respondents of Nandamalligadda are more satisfied than respondents of other villages. 50 per cent of the respondents in Nandamalligadda identified that this Agro-forestry project empowered people very high. In C.P.

Thanda and Chowderpally a majority of the respondents identified that this project had no impact on overall empowerment of villagers, i.e. 40 and 43 per cent respectively.

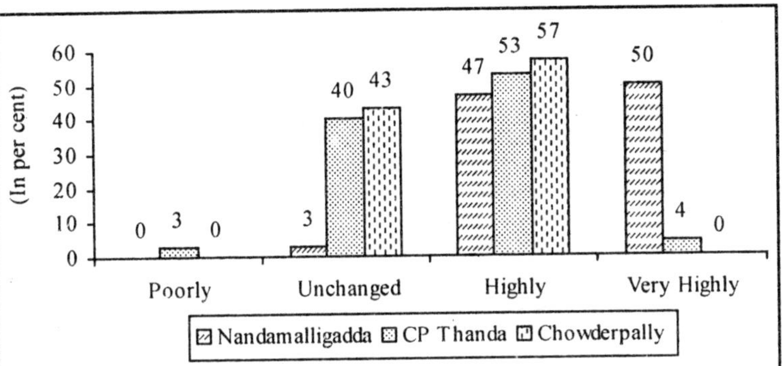

Overall Empowerment (Village-wise)

To assess the efficacy of the project with respect to overall empowerment in all the three villages, Analysis of Variance was done and the hypotheses proposed were:

Null Hypothesis—H_0: There is no significant difference in overall empowerment of the respondents of all the three villages.

Alternative Hypothesis—H_1: There is a significant difference in overall empowerment of the respondents of all the three villages.

The **probability-value** for this test is **0.0000**. Also the observed *F*-value falls in the rejection zone, at a significance level of 5 per cent. Hence, we reject the null hypothesis. Alternatively, we accept that there is a significant difference in the overall empowerment of the respondents of all the three villages. This implies that, as per the perception of the respondents there is a significant difference in the overall empowerment of the respondents of all the three villages. Alternatively, it can be inferred that *Overall Empowerment of the respondents differ from village to village.*

Extent of Expectations Met

As can be seen in the adjoining graph, only 17 per cent of the respondents responded saying that the expectations of the villagers were met very largely, whereas 12 per cent of the respondents responded saying that largely the expectations of the villagers were met, 43 per cent of the respondents responded saying that the expectations of the villagers were met moderately, 22 per cent of the

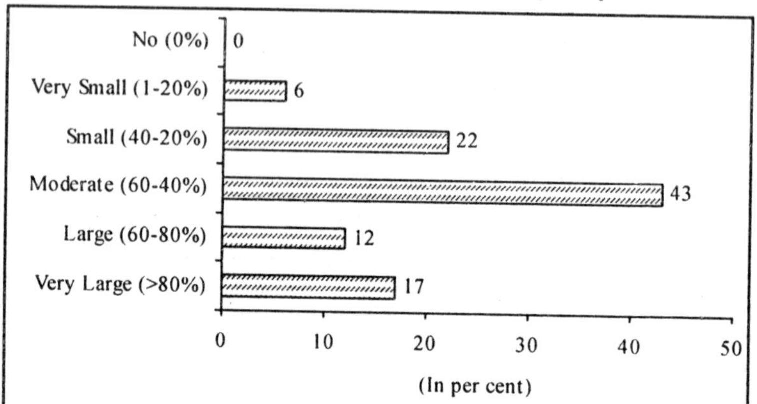

Extent to which Expectations are Met

respondents found that the extent to which the expectations of the villagers were met was small, 6 per cent of the respondents found that the extent to be very small. But none of the respondents said that the expectations of the villagers were not met . It goes on to say that, the majority of the respondent's expectations were met up to some extent.

As can be seen from adjoining graph a majority of respondents in all the villages are of the opinion that the project met their expectation moderately.

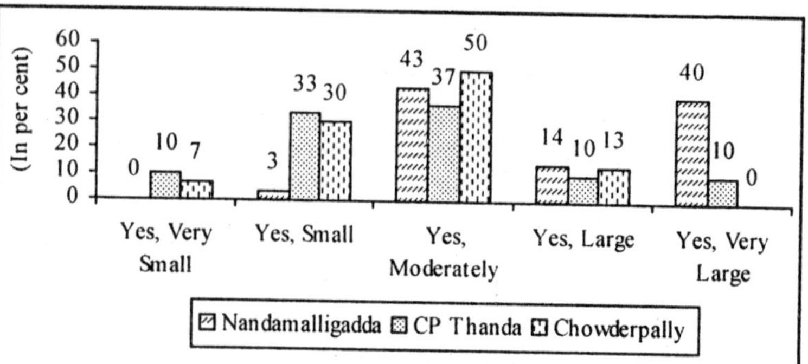

Expectations Met (Village-wise)

Respondents from Nandamalligadda gave a positive feedback about meeting expectation of people compare to respondents of other village. In C.P. Thanda and Chowderpally about 33 per cent of the respondents told that this project met their expectation only 20 to 40 per cent. All the respondents from all three villages said this project has met some part of their expectation. Not one among the respondents is completely disappointed from this agro-forestry project.

To assess the efficacy of the project with respect to meeting the expectations of the respondents of all the three villages, Analysis of Variance was done and the hypotheses proposed were:

Null Hypothesis—H_0: There is no significant difference in the level of expectations met in all the three villages.

Alternative Hypothesis—H_1: There is a significant difference in the level of expectations met in all the three villages.

The **probability-value** for this test is **0.0000**. Also the observed *F*-value falls in the rejection zone at a significance level of 5 per cent. Hence we reject the null hypothesis. Alternatively, we accept that there is a significant difference in the level of expectations met in all the three villages. This implies that, as per the perception of the respondents there is a significant difference in the level of expectations met in all the three villages. Alternatively, it can be inferred that *level of expectations met differ from village to village.*

To assess whether there exists any dependency between educational level and level of fulfilment of the respondent's expectations, Chi Square Test was done and the hypothesis proposed were:

Null Hypothesis—H_0: The fulfilment of expectations and education level are independent.

Alternative Hypothesis—H_1: Fulfilment of expectations depends on education level of the respondents.

The **probability-value** for this test is **0.272**. Also the observed value falls in the acceptance zone, at a significance level of 5 per cent. Hence we accept the null hypothesis. Alternatively, we accept that fulfilment of expectations and educational level is independent of each other. This implies that, as per the perception of the respondents fulfilment of expectations and educational level are independent. Alternatively, it can be inferred that *the level of fulfilling the expectaions of the participants is independent of the educational level of the respondents.*

Success of Interactive Bottom up Approach

As can be seen in the adjoining graph, only 6 per cent of the respondents responded saying that Interactive Bottom up approach adopted for the project was very highly successful, whereas 20 per cent of the respondents responded saying that Interactive Bottom up

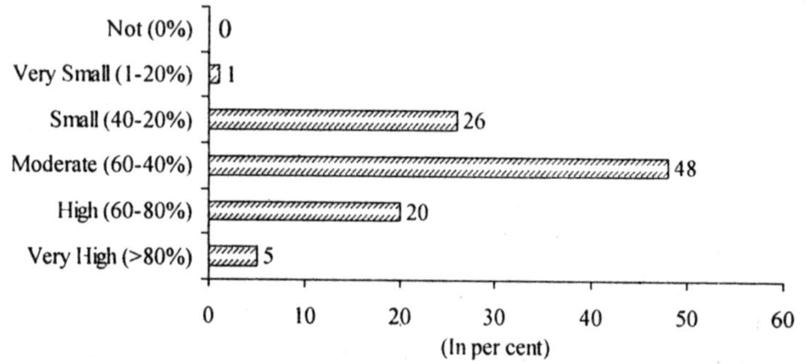

Success of Interactive Bottom up Approach

approach adopted for the project was highly successful, 47 per cent of the respondents responded saying that the success story of Interactive Bottom up approach adopted for the project was moderate, 26 per cent of the respondents found that the extent of success of Interactive Bottom up approach adopted for the project was small, 1 per cent of the respondents found the extent of success to be very small. But none of the respondents said that the Interactive Bottom up approach adopted for the project was not successful. It goes on to say that, the majority of the respondents believed that the Interactive Bottom up approach adopted for the project was successful up to an extent.

As can be seen from adjoining graph that a majority of the respondents in C.P. Thanda and Chowderpally are of the opinion that Interactive Bottom up approach is moderately successful. But Nandamalligadda is the only village where maximum respondents are of the opinion that Interactive Bottom up approach is highly successful.

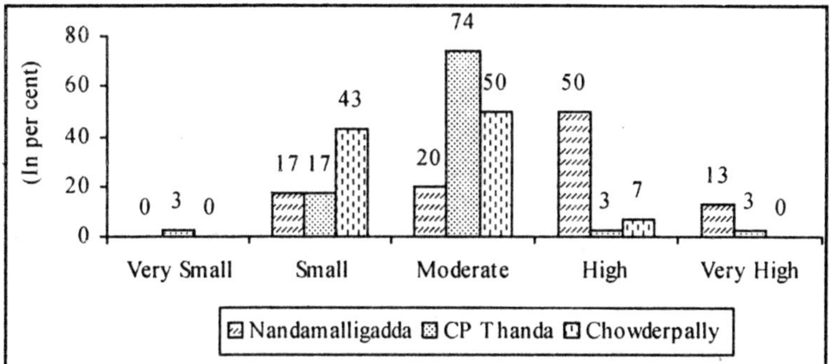

Success of Interactive Bottom Up Approach (Village-wise)

To assess the efficacy of the project with respect to the success of Interactive Bottom up approach implemented in all the three villages, Analysis of Variance was done and the hypothesis proposed were:

Null Hypothesis—H_0: There is no significant difference in the level of success of Interactive Bottom up approach in the three villages.

Alternative Hypothesis—H_1: There is a significant difference in the level of success of Interactive Bottom up approach in the three villages.

The **probability-value** for this test is **0.0000**. Also the observed *F*-value falls in the rejection zone, at a significance level of 5 per cent. Hence we reject the null hypothesis. Alternatively, we accept that there is a significant difference in the level of success of interactive bottom up approach in the three villages. This implies that, as per the perception of the respondents there is a significant difference in the level of success of interactive bottom up approach in the three villages. Alternatively, it can be inferred that *success of interactive bottom up approach differs from village to village.*

To assess whether there exists any dependency between educational level and opinion about success of interactive bottom up approach, Chi Square Test was done and the hypothesis proposed were:

Null Hypothesis—H_0: The opinion about success of interactive bottom up approach and education level are independent.

Alternative Hypothesis—H_1: The opinion about success of interactive bottom up approach and educational level are dependent.

The **probability-value** for this test is **0.509**. Also the observed value falls in the acceptance zone, at a significance level of 5 per cent. Hence we accept the null hypothesis. Alternatively, we accept that the opinion about success of interactive bottom up approach and educational level is independent. This implies that, as per the perception of the respondents the opinion about success of interactive bottom up approach and educational level is independent. Alternatively, it can be inferred that *the opinion about the success of interactive bottom up approach is independent of the educational level of the respondents.*

We can infer from this analysis that there is dependency in educational level of people and opinion of people about success of interactive bottom up approach. In other words we can say that how much ever a person is educated it will affect his opinion about success of interactive bottom up approach adopted by the project.

Overall Effectiveness of the Project

As can be seen in the adjoining graph, only 6 per cent of the respondents responded saying that the overall effectiveness of the project is very high, whereas 32 per cent of the respondents responded saying that project's overall effectiveness is high, 33 per cent of the respondents responded saying that the overall effectiveness of the project is moderate, 17 per cent of the respondents found that the overall effectiveness of the project is low, 3 per cent of the respondents found the extent of overall effectiveness of the project to be very

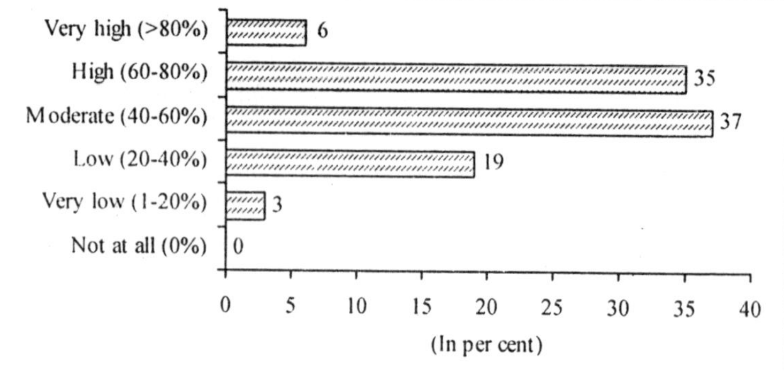

Effectiveness of the Project (Overall)

low. But none of the respondents said that the overall effectiveness of the project is nill. It goes on to say that, the majority of the respondents believed that the project is effective overall up to an extent (ranging from very large to very low).

As can be seen from the adjoining graph that maximum of the respondents gave positive response about overall effectiveness of project in Nadamalligadda and C.P. Thanda. Maximum of the respondents in Chowderpally said effectiveness of the project is moderate. All the respondents identified some effectiveness of the agro-forestry project. Chowderpally is the only village where a majority of the respondents identified project's effectiveness is moderate. Nandamalligadda and C.P. Thanda have 6.66 per cent and 40 per cent respondents who identified effectiveness of respondents is low or very low.

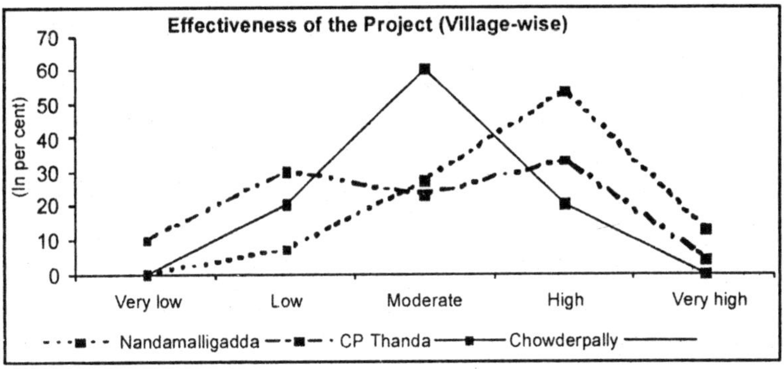

To assess whether there exists a significant difference in respondent's opinion from all the three villages about the effectiveness of agro-based forestry, Analysis of Variance was done and the hypothesis proposed were:

Null Hypothesis—H_0: There is no significant difference in the opinion of the people about the success of agro-based forestry in all the three villages.

Alternative Hypothesis—H_1: There is a significant difference in the opinion of the people about the success of agro-based forestry in all the three villages.

The **probability-value** for this test is **0.0000**. Also the observed *F*-value falls in the rejection zone, at a significance level of 5 per cent. Hence we reject the null hypothesis. Alternatively, we accept there is a significant difference in the opinion of the people about the success of agro-based forestry in all the three villages. It can be inferred that *Effectiveness of agro-based forestry project differs from village to village*.

To assess whether there exists any dependency between educational level and opinion about effectiveness of agro-forestry project, Chi Square Test was done and the hypothesis proposed were:

Null Hypothesis—H_0: The opinion about the effectiveness of agro-forestry project and education level are independent.

Alternative Hypothesis—H_1: The opinion about the effectiveness of agro-forestry project and education level are dependent.

The **probability-value** for this test is **0.181**. Also the observed value falls in the acceptance zone, at a significance level of 5 per cent. Hence we accept the null hypothesis. Alternatively, we accept that the opinion about the effectiveness of agro-forestry project and educational level is independent. This implies that, as per the perception of the respondents the opinion about the effectiveness of agro-forestry project and educational level is independent. Alternatively, it can be inferred that *the opinion about the success of agro-based forestry project is independent of the educational level of the respondents.*

Mini Case: 9

Beneficiary: Jetta Narsimha Village: Chowderpally NGO: PEACE

Brief Background Information: He is at 68, having two sons. One of them was in economically poor condition. The problem was he took loan from the bank for the borewell; the borewell provided water for one acre (two crops). His ability to feed his entire family with the support of these crops was difficult as the other part of the crop was highly dependent on rain. He has total 17 acres of land. Out of 17 acres only 2 acres is irrigated and rest is rainfed. Out of 15 acres he gave 3 acres for cultivation under agro-forestry.

Experience in Plantations: He said the mango trees provided them money from the last two years, and family members also had an opportunity to eat sufficiently. These plants didn't survive because of lack of knowledge and experience on the part of the farmer. But now they have perfect knowledge of this type of plants. He started sharing experiences with other farmers also who come to him or in village meeting. Now many farmers show interest and the crop gives him pleasure when he is visits the field.

Opinion on Field and Boundary Plantation: He said that out of 120 mango plants, now there are nearly 80 plants which gave good yield in the last two years. Cucumber, drum stick and guava trees are also many in number. He said boundary trees help in protecting the crop.

Overall Opinion: Because of problem that the farmers faced, they did not show interest in agriculture. If they had shown interest on plantation in those days itself the village would have been greener than today. According to him caste discrimination and personal disputes are strong hurdles for development of the village.

Suggestions as a Practitioner: As a practitioner of the agro-forestry cultivator he suggested to other dryland farmers to show interest on innovative way of cultivation. He accepted that farmers still believe in old methods. He expressed that he had no hope in the initial stages of the crop. But after one year he gained confidence on agro-forestry based cultivation. He believes that this system is good for other crops also.

8

Discussions

It conclusively emerges from the analyses and findings of the primary data that the project was successful in showing an impact on the participants specifically and in and around the project sites generally. More than any thing, it was able to leave social impact on the project site. This is one of the biggest achievements from the innovation diffusions of present nature which is focused more on transfer of technology than on the people and processes for whom they are meant.

Social Impact

The project was very successful in spreading the awareness about agro-ecological approach of agro-forestry at all the project sites. This can be buttressed by the fact that before the initiation of the project only four per cent of the respondents had heard about agro-forestry. Similar thing was observed with respect to the family members of the respondents too. The findings suggest that a very meagre percentage, i.e., only one per cent of the family members of the respondents had even heard about agro-forestry before the commencement of the project. In addition, the change in awareness has come as part of the project. It can be inferred from the fact that six per cent of the respondents feel that their awareness about agro-forestry has increased by one-fold; seventeen per cent of them feel that their awareness about agro-forestry has increased by two-fold; twenty nine per cent and twenty seven per cent of the respondents feel that their awareness about agro-forestry has increased by three-fold and main-fold respectively.

As far as developing the horizontal and vertical relationships and hence the increase in social capital is concerned, the project was moderately successful in doing the same. This can be gauged from the fact that a majority of respondents, i.e., sixty-eight per cent said that a moderate level of horizontal relationship has developed whereas seventeen per cent of the respondents said that a high level of horizontal relationship has developed. In addition, a majority of respondents, i.e., fifty-seven per cent said that a moderate level of horizontal relationship has developed whereas seventeen per cent of the respondents said that a high level of horizontal relationship has developed. From this, it can be conclusively inferred that social capital among the respondents have significantly increased. But further in depth studies specifically focusing on the measurement of social capital before and after the project needs to be done. This again is a very good indication of the overall success of the project.

As far as the relational changes in the participants were concerned in terms of decrease in discrimination by caste, gender and economic class, increase in the participation in village development activities, awareness of their rights, awareness about their responsibilities and saving for future, project was not successful in making much headway. This can be inferred from the fact that more than half, i.e., fifty-five per cent of the respondents accepted that there is no effect of agro-forestry project on decrease of discrimination on the basis of caste, gender and economic class. Also, seventeen per cent of the respondents agree and three per cent of respondents strongly agree that this project had a positive impact to decrease the discrimination. But twenty-five per cent of respondents disagree and one per cent of respondents strongly disagree that the project was successful in decreasing social discrimination.

In addition, forty-five per cent respondents said that the project had no effect in increasing the participation in village development. Moreover, twenty-seven per cent respondents disagreed and twenty-eight per cent of respondents strongly disagreed that this project had motivated them to participate in the village development activities. In fact, not even a single respondent was of the view that this project has resulted in increasing participation in village development activities.

Added to that, a majority of respondents feel that the project was not successful in making them aware about their rights and responsibilities. This trend is more pronounced in project's failure to make the respondents realize their responsibilities first as the beneficiaries of the project and second, as an innovative farmer and third and final, as a good citizen. An overwhelming majority of the respondents strongly disagreed with the statement that the project was able to increase the awareness regarding their responsibilities. Moreover, only a minuscule percentage of respondents felt that it did not change the awareness level related to their responsibilities.

These things are a strong reminder of the fact that any developmental project which intends to bring in qualitative change in the life of beneficiaries has to adopt a holistic approach towards its implementation. Not only should it focus on the stated objectives but it should see the overall imperative of the whole initiative. In the present case, it would have been very beneficial had the implementing authorities given equal importance to the issues of relational change, increase in social capital and overall social and qualitative changes as well and would not have focused just on the transfer of technology. This becomes much more imperative to ensure the sustainability of the whole effort. Any diffusion of innovation has to see the soft levers, i.e., the people and the processes in order to bring about holistic change and ensuring the sustainability of the project. But it was able to bring a moderate level of social and overall empowerment of the participants which is a very good sign of success of project.

In addition to this developmental projects of this nature should bring into focus the empowerment of beneficiaries as an important impact. Without participants getting empowered the essence of the initiative gets lost as soon as the project period is over. So, the empowerment of participants not only assures the holistic development but also ascertains that the effort put in for the initiative would be sustained in the long-run. The present project was moderately successful in empowering the participants socially. This can be gauged from the fact that nineteen per cent of the respondents felt that they have been very highly empowered socially after the initiation of agro-forestry project, whereas forty-five per cent of the respondents felt highly empowered socially. Similarly, as far as overall empowerment of the respondents is concerned, eighteen per cent of the respondents felt there was very high overall empowerment after the agro-forestry project, whereas fifty-two per cent of the respondents felt there was high overall empowerment after the agro-forestry project.

On the whole, the present project has been able to achieve a social impact. Specifically it was very successful in generating and spreading awareness about agro-ecological approach for the benefit of small and marginal farmers of the dryland tracts. It also has achieved success in the diffusion of technology by removing and minimizing constraints for the adoption of specific agro-forestry practices. But the project could not achieve similar success in forging relationships and increasing the social capital thereby. Also, it failed to decrease discrimination by caste, gender and economic class, increase the participation in village development activities increase awareness about their rights and responsibilities.

Initiative Replication

The successful diffusion of an innovation and its sustainability are prerequisites for any developmental initiative for replication. In fact the best approach is first to concentrate on effective diffusion to the extent that it becomes a model in itself. This instills the confidence required for replication of the initiative and adequate responses from all the stakeholders. The present initiative of promoting agro-forestry systems in the dryland tract of Mahboobnagar and Nalgonda is an ideal and innovative example of dovetailing the developmental needs of rural poor with the recent advancement in technology. Introduction of agro-forestry at the project site has helped in developing these dryland tracts based on sound agro-ecological principles. These wastelands when fully developed would provide fuelwood, fodder, fruits, fiber and oils. The small and marginal farmers adopting this system of land use have already started getting fuelwood and fodder besides providing additional income.

Also, the project has been overwhelmingly successful in bringing awareness about agro-forestry to not only the beneficiaries of the project but also the entire village and the surrounding areas. This has resulted in a melded and ready knowledge base of traditional practices and recent technological know-how at the village level. In addition to this, there were no serious constraints faced by the respondents in the implementation of agro-forestry projects. Moreover, the project has been so successful at one of the project site that it has become a model in itself. Farmers and other practitioners are coming from far-flung area to see the benefits generated by the project to this place.

In addition to this there are some other fact-based evidences which point out that the requisite conditions of replicating and upscaling the project are present. Firstly, project has been successful in meeting the expectations of the beneficiaries to a large extent. In fact, seventeen per cent of the respondents felt that their expectations were met very largely, whereas twelve per cent of the respondents pointed out that largely the expectations of the villagers were met and forty-three per cent of the respondents mentioned that their expectations were met moderately. Significantly, none of the respondents said that their expectations were not met by the project. Similarly, on the overall effectiveness front project has faired well. Majority of the respondents believe that the project has been effective towards its set objectives. This can be gathered from the finding that only six per cent of the beneficiaries responded saying that the overall effectiveness of the project is very high whereas thirty-two per cent of them felt that it was high and thirty-three per cent of the respondents termed it moderate.

Also, there is a general sense of preparedness and ready acceptance for the project of similar nature which is a crucial factor in effective upscaling of the initiative particularly so with the initiative taken up by APNLBP. This can be gauged from the fact that a majority of the beneficiaries, fifty-two per cent of them said that they are willing to promote the agro-forestry systems to a very large extent. In tune with this thirty-two per cent of the project participants said that they would promote it to a large extent. Moreover, ten per cent of the beneficiaries said that they will promote the agro-forestry project to a moderate or small extent. A similar trend is visible in the responses related to participation in extension activities related to agro-forestry activities. Again, an overwhelming majority of participants said that they would be willing to participate in the extension activities of agro-forestry. This can be inferred from the fact that a majority of the beneficiaries, fifty-two per cent of them said they are willing to participate in the extension activities pertaining to agro-forestry systems to a very large extent. In tune with this thirty-four per cent of the project participants replied that they would involve in it to a large extent. Moreover, twelve per cent of the beneficiaries said that they participate in the extension activities a moderate or small extent.

Finally, for successful replication of an initiative the intended participants should have enough confidence in the approach adopted for the diffusion and implementation of that initiative. In the absence of this, a considerable amount of time and other resources would have to be put forward. Moreover, if the approach acquires an unfavourable impression, then it becomes difficult and almost impossible to replicate the same. In the present case, the overall approach adopted by APNLBP that of interactive bottom up has got a very favourable image. This can be inferred from the fact that six per cent of the respondents felt Interactive Bottom up approach adopted for the project was very highly successful, whereas twenty per cent of the respondents felt it was highly successful, forty-seven per cent of the respondents termed the approach as moderate success whereas twenty-seven per cent of them found that the extent of success of Interactive Bottom up approach adopted for the project was small. Significantly not even a single respondent felt that the Interactive Bottom up approach adopted for the project was a failure. It goes on to say that, the majority of the respondent's believed that the Interactive Bottom up approach adopted for the project was successful upto an extent.

In light of the above points it can be conclusively said that this project can be replicated to other dryland tracts. In fact it has become a successful template for the replication for sound agro-ecological practices. The present participatory agro-forestry project for poverty alleviation and environmental restoration has allowed for the creative partnership between the different stakeholders. This model should become a template for further replication across the state and indeed the country.

Sustainable Livelihoods Approach

The Sustainable Livelihoods Framework enhances understanding of poverty and food insecurity by analyzing the relationships between relevant factors at the household, community, and regional levels. This approach explicitly examines the context in which people live in a rural community. By including concepts of vulnerability, assets, and empowerment, the Sustainable Livelihoods Framework goes beyond conventional socio-economic measures.

The adjoining figure illustrates the overall Sustainable Livelihoods Framework. The framework is dynamic and recognizes changes due to external fluctuations and the results of people's own

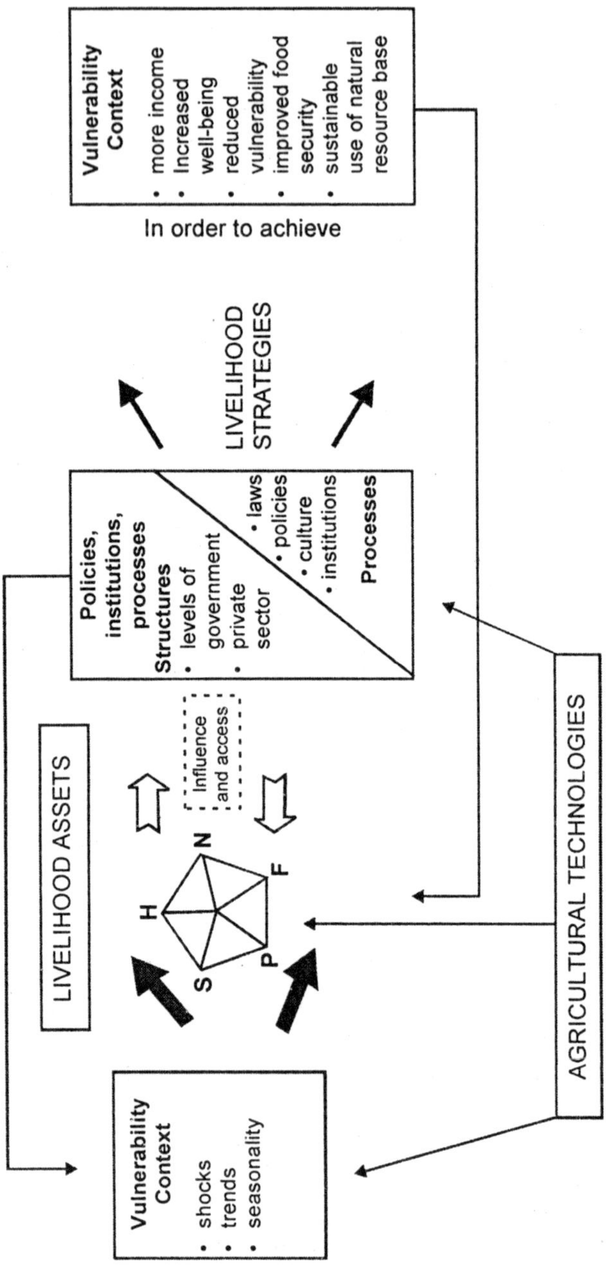

Key to "livelihood assets"; H = human; N = natural; F = financial; P = physical; S = social

actions. The starting point of the framework is the vulnerability context within which people live. The vulnerability context is affected by external influences such as the weather and price changes. External changes may also affect the assets held by people in a particular community. The framework recognizes five types of assets: human ("H" in the figure); natural (N), financial (F), physical (P), and social (S), which all need to be examined. The assets interact with policies, institutions, and processes in shaping the choice of livelihood strategies. These, in turn, shape the outcomes of livelihoods, which are the types of impacts we are interested in.

The Sustainable Livelihoods Framework draws on many considerations often excluded from agricultural economic impact studies. It is often difficult to see, however, how agricultural research and technologies may fit into the framework. The figure shows three ways for agricultural research to enter this framework: by affecting the vulnerability context, through linkages to the asset base, or as part of the policies, institutions, and processes.

Many factors may intervene in a Sustainable Livelihoods approach. Agriculture is only one part of people's livelihoods, and agricultural research and technologies affect only part of the total farming system. Understanding other factors that impinge on livelihood can be critical for improving the ultimate impact of agricultural research. To facilitate that the present study collected quantitative and qualitative data for pathways, allowing strong economic and social analysis to facilitate their evaluation.

The livelihoods approach is concerned first and foremost with people. It seeks to gain an accurate and realistic understanding of people's strengths (assets or capital endowments) and how they endeavour to convert these into positive livelihood outcomes. The approach is founded on a belief that people require a range of assets to achieve positive livelihood outcomes; no single category of assets on its own is sufficient to yield all the many and varied livelihood outcomes that people seek. This is particularly true for poor people whose access to any given category of assets tends to be very limited. As a result they have to seek ways of nurturing and combining what assets they do have in innovative ways to ensure survival. A brief discussion of the framework with its definition has been done below. This is followed by the analysis of present project on the basis of this framework.

The Vulnerability Context

The *Vulnerability Context* frames the external environment in which people exist. People's livelihoods and the wider availability of assets are fundamentally affected by critical **trends** as well as by **shocks** and **seasonality**—over which they have limited or no control.

The factors that make up the *Vulnerability Context* are important because they have a direct impact upon people's asset status and the options that are open to them in pursuit of beneficial livelihood outcomes.

- *Shocks* can destroy assets directly (in the case of floods, storms, civil conflict, etc.). They can also force people to abandon their home areas and dispose of assets (such as land) prematurely as part of coping strategies. Recent events have highlighted the impact that international economic shocks, including rapid changes in exchange rates and terms of trade, can have on the very poor.
- *Trends* may (or may not) be more benign, though they are more predictable. They have a particularly important influence on rates of return (economic or otherwise) to chosen livelihood strategies.
- *Seasonal Shifts* in prices, employment opportunities and food availability are one of the greatest and most enduring sources of hardship for poor people in developing countries.

Human Capital

Human capital represents the skills, knowledge, ability to labour and good health that together enable people to pursue different livelihood strategies and achieve their livelihood objectives. Human capital appears in the generic framework as a livelihood asset, that is, as a building block or means of achieving livelihood outcomes. Its accumulation can also be an end in itself.

Social Capital

There is much debate about what exactly is meant by the term 'social capital'. In the context of the sustainable livelihoods framework it is taken to mean the social resources upon which people draw in pursuit of their livelihood objectives. These are developed through:

- *networks and connectedness*, either vertical (patron/client) or horizontal (between individuals with shared interests) that increase people's trust and ability to work together and expand their access to wider institutions, such as political or civic bodies;
- *membership of more formalised groups* which often entails adherence to mutually-agreed or commonly accepted *rules, norms and sanctions;* and
- *relationships of trust, reciprocity and exchanges* that facilitate co-operation, reduce transaction costs and may provide the basis for *informal safety nets* amongst the poor.

The above are all inter-related. For example, membership of groups and associations can extend people's access to and influence over other institutions. Likewise trust is likely to develop between people who are connected through kinship relations or otherwise. Of all the five livelihood building blocks, social capital is the most intimately connected to *Transforming Structures and Processes*. In fact, it can be useful to think of social capital as a product of these structures and processes, though this over-simplifies the relationship. Structures and processes might themselves be products of social capital; the relationship goes two ways and can be self-reinforcing. For example:

- When people are already linked through common norms and sanctions they may be more likely to form new organisations to pursue their interests;
- Mutual trust and reciprocity lower the costs of working together. This means that social capital has a direct impact upon other types of capital;
- By improving the efficiency of economic relations, social capital can help increase people's incomes and rates of saving (financial capital); and
- Social networks facilitate innovation, the development of knowledge and sharing of that knowledge. There is, therefore, a close relationship between social and human capital.

Natural Capital

Natural capital is the term used for the natural resource stocks from which resource flows and services (e.g. nutrient cycling, erosion protection) useful for livelihoods are derived. There is a wide variation in the resources that make up natural capital, from intangible public goods such as the atmosphere and biodiversity to divisible assets used directly for production (trees, land, etc.).

Within the sustainable livelihoods framework, the relationship between natural capital and the *Vulnerability Context* is particularly close. Many of the shocks that devastate the livelihoods of the poor are themselves natural processes that destroy natural capital (e.g. fires that destroy forests, floods and earthquakes that destroy agricultural land) and seasonality is largely due to changes in the value or productivity of natural capital over the year.

Clearly, natural capital is very important to those who derive all or part of their livelihoods from resource-based. However, its importance goes way beyond this. None of us would survive without the help of key environmental services and food produced from natural capital. Health (human capital) will tend to suffer in areas where air quality is poor as a result of industrial activities or natural disasters (e.g. forest fires).

Physical Capital

Physical capital comprises the basic infrastructure and producer goods needed to support livelihoods.

- Infrastructure consists of changes to the physical environment that help people to meet their basic needs and to be more productive.
- Producer goods are the tools and equipment that people use to function more productively.

Financial Capital

Financial capital denotes the financial resources that people use to achieve their livelihood objectives. The definition used here is not economically robust in that it includes flows as well as stocks and it can contribute to consumption as well as production. However, it has been adopted to try to capture an important livelihood building block, namely the availability of cash or equivalent that enables people to adopt different livelihood strategies.

There are two main sources of financial capital:

- *Available Stocks*: Savings are the preferred type of financial capital because they do not have liabilities attached and usually do not entail reliance on others.
- *Regular Inflows of Money*: Excluding earned income, the most common types of inflows are pensions, or other transfers from the state, and remittances.

In order to make a positive contribution to financial capital these inflows must be reliable. Financial capital is probably the most versatile of the five categories of assets.

- It can be converted—with varying degrees of ease, depending upon *Transforming Structures and Processes*—into other types of capital.
- It can be used for direct achievement of livelihood outcomes—for example when food is purchased to reduce food insecurity.
- Rightly or wrongly, it can also be transformed into political influence and can free people up for more active participation in organisations that formulate policy and legislation and govern access to resources.

However, it is also the asset that tends to be the least available to the poor. Indeed, it is because the poor lack financial capital that other types of capital are so important to them.

There are, in addition, assets or desirable outcomes that may not be achievable through the medium of money.

Transforming Structures and Processes

Within the livelihoods framework are the institutions, organisations, policies and legislation that shape livelihoods. Their importance cannot be over-emphasised. They operate at all levels, from the household to the international arena, and in all spheres, from the most private to the most public. They effectively determine:

- *access* (to various types of capital, to livelihood strategies and to decision-making bodies and sources of influence);
- the *terms of exchange* between different types of capital; and
- *returns* (economic and otherwise) to any given livelihood strategy.

In addition, they have a direct impact upon whether people are able to achieve a feeling of inclusion and well-being. Because culture is included in this area they also account for other 'unexplained' differences in the 'way things are done' in different societies. The influence of *Transforming Structures and Processes* extends throughout the framework.

Livelihood Outcomes

Livelihood Outcomes are the achievements or outputs of *Livelihood Strategies*. Once again, the important idea associated with this component of the framework is that we, as outsiders, investigate, observe and listen, rather than jumping to quick conclusions or making hasty judgements about the exact nature of the outcomes that people pursue. In particular, we should not assume that people are entirely dedicated to maximising their income. Rather, we should recognise and seek to understand the richness of potential livelihood goals. This, in turn, will help us to understand people's priorities, why they do what they do, and where the major constraints lie.

The present project has tried to reduce the vulnerability context by focusing mainly on increase in Human Capital. It has done so by organizing training programmes creating awareness about the agro-forestry project and in turn trying to prevent the effect of seasonality in terms of poor or no rain by application of agro-forestry systems. As discussed in the previous section, the project has been very successful in increasing the human capital by a participatory approach. In fact the traditional knowledge has been melded effectively with the modern technological know-how and has resulted in a community-based knowledge pool which consists of skills specific to the application of agro-forestry.

Although the project focused on increasing the social capital by making it mandatory to have community-based organization implements it with the help of NGOs, it was not that successful in doing so. Only in one of the study area, it was successful in significantly increasing the knowledge base, viz., Nandamalligadda, and it has resulted in becoming a model village. This stresses the fact that any project of similar nature should focus more on increase in the social capital along with human capital for the long-term sustainability of the project. Had the NGOs of other villages also focused on this particular aspect they would have emerged very successful in the implementation and long-term effect of the project.

The project also focused on increasing the natural capital since it was using agro-ecological approach to poverty alleviation. It goes on to say that the initiative had been very successful in increasing the natural capital thereby ensuring that the natural resource base is increased. More specifically, it has resulted in good tree density crown cover and species richness. As a consequence to this, there is ready availability of fodder and fuelwood which in turn is ensuring that now there is low pressure on the natural stock in the form of nearby forest areas.

The project was not focused on increase of the physical capital as its objective was to help dryland farmers in gaining additional income using an agro-ecological approach hence there were no specific activities that were taken up related to this.

As mentioned earlier, one of the objectives was to increase the financial capital by helping add to the existing income of the farmer. The project has been moderately successful in doing so as the farmers are getting additional income from the agro-forestry system that they have adopted.

So, overall it can be said that the project has become successful in decreasing the vulnerability context of the participants by the innovative use of recent advances in agricultural technology through increasing human, social and natural capital which have in turn resulted in the increase of financial capital.

9

Conclusion and Recommendations

Recent advances in agricultural applications of modern biotechnology show a significant potential of agricultural biotechnology to contribute to sustainable gains in agricultural productivity, reducing poverty, and enhancing food security. However, there were many pathways through which the rural poor can obtain benefits or incur costs arising from agricultural research (Kerr and Kolavalli, 1999). The authors note that "technology's role in alleviating poverty is both indirect and partial; technology alone cannot overcome poverty, nor can continued poverty be blamed on improved technology."

Keeping this in consideration, it can be concluded that the Network Project on Agro-forestry has been able to achieve a moderate success in keeping with its stated objective and prime focus. It has followed a participatory bottom up approach which is the hallmark of entire initiative. It has been able to generate a great deal of awareness among the small and marginal farmers of dry hinterland regarding agro-ecological approach towards livelihoods. The project made sure that there remain no serious constraints in the adoption process of the technology. It was able to help provide additional income; decrease migration to a moderate level; increase the agricultural productivity; increase self-development opportunities and helped in village development. Also, it helped bring biological sustainability to the project site by highly increasing the tree density, richness of species, crown cover, forest product availability, water availability and decrease in soil erosion. The project has been successful in meeting the expectations of the beneficiaries to a large extent. It has been able to bring forth a general sense of preparedness and ready acceptance for the project of similar nature. Moreover, it has resulted in social and overall empowerment of the participants. This goes on to say that the present project can become a successful model worthy of replication on other dryland tracts following a integrative and congruent agro-ecological approach.

However, there always is scope of refinement in any endeavour. The present project is not immune to the same. The innovative application can become more fruitful if it is able to incorporate recommendation put forth by the study team after an indepth study both academic as well as field-based. These recommendations have been discussed briefly below.

Climate change mitigation through land use is a growing trend. The convergence of global environmental priorities with national-level stakeholders' interests is allowing the rapid development of a new carbon economy. 'New carbon economy' refers to the series of market-based policy instruments designed to reduce global greenhouse gas emissions through the creation of markets for carbon. In the global arena, for example, the Clean Development Mechanism of the Kyoto Protocol and the World Bank's Carbon Funds share the aim to promote investment in energy or forestry-related projects in developing countries. Involvement of a wide range of stakeholders in these developments is increasing, and numbers of projects are growing around the world. In light of this, it is recommended that,

The agro-forestry project should explore further linkages and possible integration of carbon-trading as a secondary benefit to the participants to ensure sustainability of the effort.

Social capital is a concept that can be utilized in any delivery approaches for the efficient transfer of agricultural technologies. Over the decades, technology adoption and diffusion in agriculture have always been a problem with no leap forward solutions in sight. Human behaviour, however, is the result of interactions and interrelations between people. Thus, using this concept enhances any approach to understanding technology adoption and diffusion in that it makes it more unified, holistic, and comprehensive by looking beyond the individual at the individual in relation with another individual. It looks at the norms and values operating in social relations and it looks at the ethos of social relations. Since, there are a large number of technologies that are being developed and implemented by APNLBP, with this regard, it is recommended that,

A study mainly focusing on developing a framework for measurement of social capital at the project site should be taken up

to understand the changes in social capital as effected by the specific project.

One of the difficulties that was faced during the conduct of this study was that there was lack of baseline information of participants and the project site. As is evident, in the absence of baseline data it becomes very difficult to gauge the impact of any initiative by following a pre-post impact assessment because the recall value of the data gathered from the participants is not that reliable. So, for a conclusive, robust and objective conduct of study hereafter, it is suggested that,

A baseline study of the participants and the project site should be made at the time of initiation of any project.

Also, there are no pre-defined criteria and indicators for monitoring evaluation and impact assessment of the projects undertaken. It is, therefore, further suggested here that,

For the projects that are taken up a log-frame including the criteria and indicators should be in place for effective monitoring, evaluation and impact assessment.

It was found from the study that on the front of formation and nurturing of CBOs, enough effort was not put forth which is very crucial for sustaining the initiative. Hence, it is recommended that,

A concerted and whole-hearted effort should go in the formation, nurturing and developing of CBOs for ensuring the sustainability of the project.

In addition, it also emerged from the study that there were no robust criteria laid down for the selection of beneficiaries, NGOs and the project site. It is, therefore, recommended here that,

A robust set of criteria should be laid down and strictly adhered to for the selection of beneficiaries, NGOs and the project sites.

Also, it conclusively emerged from the study that the project lacked a strong pro-poor bias which should have been ingrained in the initiatives. Hence it is suggested that,

The project should adopt a pro-poor bias in its approach and make sure that the benefits of the efforts are channeled to poor and the poorest of poor.

In addition, it was also felt that there were not enough efforts put in for adapting the training programmes from the perspective of farmers. Hence it is suggested that,

The project should have a robust training programme adapted to the needs and cultural settings of the intended beneficiaries.

Finally, it was observed that project was not successful in making relational change in the social setting of the participants as it focused more on technology-transfer approach rather than a holistic approach. It is, therefore, recommended that,

The project should put in enough efforts and direct its energies not only to technology-transfer but to the overall empowerment and upliftment of the beneficiaries by adopting a holistic approach.

Data Collection Tools Used

QUESTIONNAIRE FOR BENEFICIARIES

Impact of Biotechnology at Micro Level
Network Project on Agro-forestry, APNLBP

Respondent Number: [ ] **Hamlet:** [ ]

Village [ ] [ ] [ ]

Mandal [ ] [ ] [ ]

District Mehboobnagar | Nalgonda

Implementing NGO [ ]

Name of Enumerator [ ]

Date of Survey [ ]

Data Editing on [ ] **Data Feeding on** [ ]

Institute of Public Enterprise
OU Campus, Hyderabad, Andhra Pradesh

A. Background Information

1. Name of Respondent: **2. Age:**

3. Sex

1. Male	2. Female

4. Caste

1 General	2 OBC	3 Scheduled Caste	4 Scheduled Tribe

5. Education Level

1. Illiterate	2. Just literate	3. Primary	4. Middle	5. Secondary	6. Higher Sec	7. Graduate and above

6. Main Occupation

1. No work	2. Wage Laborer	3. Share cropper	4. Farmer	5. Service	6. Business

7. HH size*

1. Male Child	2. Male Adult	3. Female Child	4. Female Adult

* Write the number under each Category

8. Details of HH Members* *(Excluding the Respondent)*

* Enter the respective code only

S. No.	*Category (For code, refer Q. No. 7)*	*Education Level (For code, refer Q. No. 5)*	*Employment Status (Code: Dependent-1, Employed-2)*
1			
2			
3			
4			
5			
6			
7			
8			
9			
10			
11			
12			

9. HH Type

1. *Kuchcha*	2. *Pucca*

10. Assets/Infrastructure

1. Bicycle	2. Bullock cart	3. Radio	4. Television	5. Electricity	6. Tubewell/Well

11. Number of Cattle Owned

1.Goat =	2. Sheep =	3. Cow and Buffalo =

12. Land Holding (*In acres*): Extent of land under the project

1. Irrigated =	2. Rainfed =

13. Average annual HH income prior to the beginning of the project

1. Less than 2000	2. 2000-6000	3. 6000-10000	4. 10000-15000	5. Above 15,000

B. Awareness about Agro-forestry

1. **Did you know about Agro-forestry before the beginning of this project?***

   (Tick in the box against the appropriate response)

   1. Did not know anything about Agro-forestry ☐
   2. Knew a bit, had some idea ☐
   3. Knew, but did not have the technical know-how ☐
   4. Knew, had the technical know-how, but did not have the proper guidance to practice it ☐
   5. Knew, had the technical know-how, guidance and was practicing it ☐

2. **Did any of your family members know about Agro-forestry before the beginning of this project?***

   * *(Write the total number against the appropriate response)*

   1. Did not know anything about Agro-forestry ☐
   2. Knew a bit, had some idea ☐
   3. Knew, but did not have the technical know-how ☐
   4. Knew, had the technical know-how, but did not have the proper guidance to practice it ☐
   5. Knew, had the technical know-how, guidance and was practicing it ☐

3. Where from did you get to know about Agro-forestry first?*

**(Tick in the box against the appropriate response)*

1. From some other practicing farmers. ☐
2. From radio/television programmes. ☐
3. From Krishi Vigyan Kendra as part of Government sponsored programme. ☐
4. From some other project by private agency before this project. ☐
5. From associating with this project. (Network Project on Agro-forestry) ☐

4. How did you come to know about this project?*

**(Tick in the box against the appropriate response)*

1. From my family members. ☐
2. From friends and relatives. ☐
3. From my caste people. ☐
4. From other villagers. ☐
5. From the village level functionaries. ☐
6. From the NGO staff implementing this project. ☐

5. Why did you choose to associate with this project?*

** (Tick in the box against the appropriate response)*

1. Since my family members told me to associate with it, I went for it. ☐
2. Since my friends and relatives were associating, I also went for it. ☐
3. Since my caste people were associating, I also went for it. ☐
4. Since other villagers were associating, I also went for it. ☐
5. Since the village head and other functionaries directed me to associate, I went for it. ☐
6. Since the NGO staff directed me to associate, I went for it. ☐
7. I understood the intended benefits of it, so I went for it. ☐

6. What were your expectations from this project?*

** (Tick in the box against the appropriate response)*

1. It will result in the increase of income. ☐
2. It will generate employment round the year. ☐

3. It will increase the productivity of my land. ☐
4. It will enhance my social status. ☐
5. It will stop migration. ☐

8. After associating with this project, your awareness regarding Agro-forestry:*

* *(Tick in the box against the appropriate response)*

1. Has remained at the same level? ☐
2. Has increased one fold? ☐
3. Has increased two folds? ☐
4. Has increased three folds? ☐
5. Has increased manifolds? ☐
6. Has increased to the maximum extent possible? ☐

9. After this project, the awareness regarding Agro-forestry in your village:*

* *(Tick in the box against the appropriate response)*

1. Has remained at the same level? ☐
2. Has increased one fold? ☐
3. Has increased two folds? ☐
4. Has increased three folds? ☐
5. Has increased manifolds? ☐
6. Has increased to the maximum extent possible? ☐

C. Adoption Pattern

1. Why did you adopt Agro-forestry?*

**(Rank the responses in the order of their importance)*

1. Increase in soil fertility. ☐
2. Increase in the productivity of land. ☐
3. Decrease in the soil erosion. ☐
4. Availability of fuel wood and firewood. ☐
5. Availability of fodder. ☐
6. Increase in income. ☐

2. Was the planting material given to you free of cost? If No, what was the cost of planting material?

☐

3. Have you used any biopesticides in raising the crops?

☐

4. How many grams of Farm Yard Manure did you apply per pit?

5. Please rank the three most important problems you have faced after the adoption of Agro-forestry?

1.
2.
3.

6. To what extent did the Scientists/NGO who promoted it helped you in finding solutions to the problems you faced?*

* *(Tick in the box against the appropriate response)*

1. Did not help at all	2. Helped some times	3. Helped several times	4. Helped most of the times	5. Helped all the times

7. Please indicate the severity of following constraints that you faced*

* *(Tick in the box against the appropriate response)*

Ratings	1 Not at All	2 Low	3 Moderate	4 Severe	5 Very Severe
1. Lack of Technical Information					
2. Supply of Seeds					
3. Supply of Fertilizers					
4. Others					

D. Social Impact

1. What are the support did you get from the implementing NGO?

Code	*Support*	*Received (Code: 1 = Yes; 2 = No)*	*Adequacy (Code: 5 = most; 1 = least)*
1	Funds for implementing the project		
2	Planting material and fertilisers		
2	Guidance and support		
3	Monitoring and supervision		
4	Training opportunities		
5	Technical inputs		
6	Others (Please specify)		

Formation of Community based Organisation:

1. Did you form an organisation with the help of implementing NGO?	1. Yes	2. No	3. Don't know

2. If yes, did this organisation have a committee for taking decisions on your behalf?	1. Yes	2. No	3. Don't know

3. How did the members chosen for this committee?	1. By Election	2. By Selection	3. By Consensus among members

4. Did your organisation formulate rules and regulations for the effective implementation of the project?	1. Yes	2. No	3. Don't know

5. Did the members meet for discussing about the issues related to the project?	1. Yes	2. No	3. Don't know

6. If yes, what is the frequency of the meeting?	1. Weekly	2. Half-monthly	3. Monthly	4. 3-monthly	5. 6-monthly	6. Yearly

1. How often do you have conflicts related to group matter	5. Never	4. Seldom	3. Frequently	2. Every meetings	1 Very high
How do you resolve conflict?	5. Participatory manner	4. Some member always resolve	3. Leader resolves	2. External bodies	1 No idea
How frequently conflicts are resolved?	1 Very same meeting	2. Next meeting	3. As and when time is found	4. Never resolved	5 Others

Have any encroachment/trespassing taken place after the initiation of project?	1. Yes	2. No	3. Don't know

If yes, what was the step taken	1. None	2. Your group interferes	3. Panchayat interference	4. Legal action	5 Don't know

Do you charge any penalties for the offenders?	5. Most of (Above 80%)	4. Majority (60-80%)	3. Moderately (60-40%)	2. Least (40-20%)	1. Very few (Less than 20%)
Had you ever heard about penalty/royalty amount recovered from offenders?	5. Most of (Above 80%)	4. Majority (60-80%)	3. Moderately (60-40%)	2. Least (40-20%)	1. Very few (Less than 20%)

Has your committee had established any relationship with any other organizations besides the associated NGO?	1. Yes	2. No	3. Don't know

If yes, please provide following details

Organization (Code: 1. Govt.; 2. Semi govt.; 3. Non-govt.; 4. Village based)	**Activities** (Code: 1. Financing; 2. Capacity-building; 3. Implementation; 4. Information sharing; 5. Monitoring and evaluation)	**Resources shared** (Code: 1. Human; 2. Financial; 3. Materaial; 4. None)	**Co-ordination** (Code: 1. High; 2. Moderate; 3. Low

Networking

1. Have you developed relationship among the people of your level	1. No	2. Low	3. Moderate	4. High	5. Very High
2. Have you developed relationship among the people of higher/lower level than you	1. No	2. Low	3. Moderate	4. High	5. Very High

Access to Services

1. What are the benefits you get from the Agro-forestry system?

Benefits	*Code*	*Benefits derived*	*Domestic requirement fulfillment*
Fuelwood	1		
Fodder Timber	2 3		
Non-timber forest product Employment and wage	4 5		
Environment related (Health and hygiene) Community development	6 7		
Income generation	8		

Benefit derived code: 1. Yes 2. No

Domestic requirement code: 5 if fully fulfilled and 1 if least fulfilled

2. Is there any discrimination in delivering and sharing of services and resources by?

1. Associated members themselves	1. No	2. Low	3. Moderate	4. High	5. Very High
2. Supporting NGOs	1. No	2. Low	3. Moderate	4. High	5. Very High
3. Others	1. No	2. Low	3. Moderate	4. High	5. Very High

Material change

1. HH income	5. Highly increased	4. Increased	3. No change	2. Decreased	1 Highly decreased
2. Work load	5. Highly increased	4. Increased	3. No change	2. Decreased	1 Highly decreased
3. Migration	5 Highly increased	4. Increased	3. No change	2. Decreased	1 Highly decreased
5. Fuelwood availability	5 Highly increased	4. Increased	3 No change	2 Decreased	1 Highly decreased
6. Fodder availability	5 Highly increased	4. Increased	3. No change	2. Decreased	1 Highly decreased
7. NTFP availability	5 Highly increased	4 Increased	3' No change	2 Decreased	1 Highly decreased
8. Off farm activity	5 Highly increased	4 Increased	3 No change	2 Decreased	1 Highly decreased
9. Agricultural productivity	5. Highly increased	4 Increased	3. No change	2. Decreased	1 Highly decreased
10. Self-development Opportunities	5. Highly increased	4. Increased	3. No change	2. Decreased	1 Highly decreased
11. Village Development	5. Highly increased	4. Increased	3. No change	2. Decreased	1 Highly decreased

Relational Change

1. Discrimination by caste, gender and economic class had decreased	5. Strongly agree	4. Agree	3. Don't change	2. Disagree	1. Strongly disagree
2. Participation in village development activities has increased	5. Strongly agree	4. Agree	3. Don't change	2. Disagree	1. Strongly disagree
3. People have become aware of their rights	5. Strongly agree	4. Agree	3. Don't change	2. Disagree	1. Strongly disagree
4. People have become aware of their responsibility	5. Strongly agree	4. Agree	3. Don't change	2. Disagree	1. Strongly disagree
5. People have started saving for future	5. Strongly agree	4. Agree	3. Don't change	2. Disagree	1 Strongly disagree

Biological Sustainability

After the Agro-forestry project in your village:

1. Species richness	5. Highly increased	4. Increased	3. Similar	2. Decreased	1. Highly decreased
2. Tree density	5. Highly increased	4. Increased	3. Similar	2. Decreased	1. Highly decreased
3. Crown cover	5. Highly increased	4. Increased	3. Similar	2. Decreased	1. Highly decreased
4. Plantation	5. Highly increased	4. Increased	3. Similar	2. Decreased	1. Highly decreased
5. Water	5. Highly increased	4. Increased	3 Similar	2. Decreased	1. Highly decreased
6. Soil erosion	5. Highly increased	4. Increased	3. Similar	2. Decreased	1 Highly decreased
7. Forest product availability	5. Highly increased	4 Increased	3. Similar	2. Decreased	1. Highly decreased

Economic Impact

For what purpose did you use the increase in your income?	1. Consumption	2. Production	3. Don't know

Perceptions and Feedback of Primary Stakeholder

1. Will you promote agro-forestry systems?	5. Yes, very large (> 80%)	4. Yes, large (60-80%)	3. Yes, moderate (60-40%)	2. Yes, small (40-20%)	1. Yes, very small (1-20%)	0. No (0%)
2. Will you participate in the extension activities of agro-forestry?	5. Yes, very large (> 80%)	4. Yes, large (60-80%)	3. Yes, moderate (60-40%)	2. Yes, small (40-20%)	1. Yes, very small (1-20%)	0. No (0%)

3. Did the villagers had opportunity to participate in trainings organized by associated NGO?	5. Most of (> 80%)	4. Majority (60-80%)	3. Moderately (60-40%)	2. Least (40-20%)	1. Very few (Less than 20%)
4. How are those training programmes for capacity development?	5. Very useful	4. Useful	3. No change	4. Not useful	1. Useless
5. Has there been any skill development in you?	5. Very good	4. Good	3. Satisfactory	4. Poor	1. Very Poor

6 Has there been any knowledge improvement in you?	5. Very good	4 Good	3. Satisfactory	4. Poor	1 Very Poor
7. Please rate the performance of the NGO	5. Very good	4. Good	3 Satisfactory	4. Bad	1 Very Bad

8. How do you perceive your empowerment after the initiation of Agro-forestry project?

a. Social Empowerment	5. Very Highly	4. Highly	3. Unchanged	2. Poorly	1. Very Poorly
b. Economic Empowerment	5. Very Highly	4. Highly	3. Unchanged	2. Poorly	1. Very Poorly
c. Overall Empowerment	5. Very Highly	4. Highly	3 Unchanged	2. Poorly	1. Very Poorly

9. To what extent did this project meet your expectations?	5. Yes, very large (> 80%)	4. Yes, large (60-80%)	3. Yes, moderate (60-40%)	2. Yes, small (40-20%)	1. Yes, Very small (1-20%)	0. No (0%)
10. In your opinion, the Interactive Bottom up approach adopted by the project was successful?	5. Very high (> 80%)	4 High (60-80%)	3. Moderate (60-40%)	2. Small (40-20%)	1 Very small (1-20%)	0. Not (0%)
11. How do you perceive the overall effectiveness of the project?	5 Very high (> 80%)	4. High (60-80%)	3. Moderate (60-40%)	2. Low (40-20%)	1. Very Low (1-20%)	0. Not at all (0%)

Economic Analysis

1. Main Crop

Cost of Inputs	
1. Primary Tillage	
2. Cost of Planting Material	
3. Cost of Farm Yard Manure/Fertilizers	
4. Cost of Pesticides	
5. Weeding Costs (twice)	
6. Community Watering Costs	
7. Electricity Charges	
8. Total Labour costs for all Operations	
Total Input Costs for the Year	

Output	
1. Total yield/acre (in quintals or bags or kgs)	
2. Selling price (in quintals or bags or kgs)	
Total Income	

2. Agro-forestry Crop 1(**Please specify**):
No. of Plant/acre:

Cost of Inputs	1st Year	2nd Year	3rd Year	4th Year	5th Year
1. Primary Tillage					
2. Cost of Planting Material					
3. Cost of Farm Yard Manure/Fertilizers					
4. Cost of Pesticides					
5. Weeding Costs (twice)					
6. Community Watering Costs					
7. Electricity Charges					
8. Total Labour Costs for all Operations					
Total Input Costs for the Year					

Output					
1. Average yield/plant (in quintals or kgs)					
2. Selling price (in quintals or bags or kgs)					
Total Income					

3. Agro-forestry Crop 2 (**Please specify**):
No. of Plant/acre:

Cost of Inputs	1st Year	2nd Year	3rd Year	4th Year	5th Year
1. Primary Tillage					
2. Cost of Planting Material					
3. Cost of Farm Yard Manure/Fertilizers					
4. Cost of Pesticides					
5. Weeding Costs (twice)					
6. Community Watering Costs					
7. Electricity Charges					
8. Total Labour Costs for all Operations					
Total Input Costs for the Year					

Output					
1. Average yield/plant (in quintals or kgs)					
2. Selling price (in quintals or bags or kgs)					
Total Income					

4. Agro-forestry Crop 3 (**Please specify**):
No. of Plant/acre:

Cost of Inputs	1st Year	2nd Year	3rd Year	4th Year	5th Year
1. Primary Tillage					
2. Cost of Planting Material					
3. Cost of Farm Yard Manure/Fertilizers					
4. Cost of Pesticides					
5. Weeding Costs (twice)					
6. Community Watering Costs					
7. Electricity Charges					
8. Total Labour Costs for all Operations					
Total Input Costs for the Year					

Output					
1. Average yield/plant (in quintals or kgs)					
2. Selling price (in quintals or bags or kgs)					
Total Income					

5. Agro-forestry Crop 4 **(Please specify):**
   No. of Plant/acre:

Cost of Inputs	1st Year	2nd Year	3rd Year	4th Year	5th Year
1. Primary Tillage					
2. Cost of Planting Material					
3. Cost of Farm Yard Manure/Fertilizers					
4. Cost of Pesticides					
5. Weeding Costs (twice)					
6. Community Watering Costs					
7. Electricity Charges					
8. Total Labour Costs for all Operations					
Total Input Costs for the Year					

Output					
1. Average yield/plant (in quintals or kgs)					
2. Selling price (in quintals or bags or kgs)					
Total Income					

6. Agro-forestry Crop 5 **(Please specify):**
   No. of Plant/acre:

Cost of Inputs	1st Year	2nd Year	3rd Year	4th Year	5th Year
1. Primary Tillage					
2. Cost of Planting Material					
3. Cost of Farm Yard Manure/Fertilizers					
4. Cost of Pesticides					
5. Weeding Costs (twice)					
6. Community Watering Costs					
7. Electricity Charges					
8. Total Labour Costs for all Operations					
Total Input Costs for the Year					

Output					
1. Average yield/plant (in quintals or kgs)					
2. Selling price (in quintals or bags or kgs)					
Total Income					

QUESTIONNAIRE FOR NON-BENEFICIARIES

Impact of Biotechnology at Micro Level
Network Project on Agro-forestry, APNLBP

Respondent Number: [ ] **Hamlet:** [ ]

Village [ ] [ ] [ ]

Mandal [ ] [ ] [ ]

District | Mehboobnagar | Nalgonda |

Implementing NGO [ ]

Name of Enumerator [ ]

Date of Survey [ ]

Data Editing on [ ] **Data Feeding on** [ ]

Institute of Public Enterprise
OU Campus, Hyderabad, Andhra Pradesh

A. Background Information

1. Name of Respondent: **2. Age:**

3. Sex

1. Male	2. Female

4. Caste

1. General	2 OBC	3 Schedule Caste	4 Schedule Tribe

5. Education Level

1. Illiterate	2. Just literate	3. Primary	4. Middle	5. Secondary	6. Higher Sec	7. Graduate and above

6. Main Occupation

1. No work	2. Wage Labourer	3. Share Cropper	4. Farmer	5. Service	6. Business

7. HH Size*

1. Male Child	2. Male Adult	3. Female Child	4. Female Adult

* Write the number under each category

8. Details of HH Members* *(Excluding the Respondent)*

* Enter the respective code only

S. No.	*Category (For code, refer Q. No. 7)*	*Education Level (For code, refer Q. No. 5)*	*Employment Status (Code: Dependent-1, Employed-2)*
1			
2			
3			
4			
5			
6			
7			
8			
9			
10			
11			
12			

9. HH Type

1. *Kuchcha*	2. *Pucca*

10. Assets/Infrastructure

1. Bicycle	2. Bullock cart	3. Radio	4. Television	5. Electricity	6. Tubewell/Well

11. Number of Cattle Owned

1.Goat =	2. Sheep =	3. Cow and Buffalo =

12. Land holding (*In acres*) **Extent of land under the project**

1. Irrigated =	2. Rainfed =

13. Average annual HH income prior to the beginning of the project

1. Less than 2000	2. 2000-6000	3. 6000-10000	4. 10000-15000	5. Above 15,000

B. Awareness about Agro-forestry

1. Why did you not join the Agro-forestry project? Please give three most important reasons.

1.
2.
3.

2. Did you know about Agroforestry before the beginning of this project?*

* *(Tick in the box against the appropriate response)*

1. Did not know anything about Agro-forestry. ☐
2. Knew a bit, had some idea. ☐
3. Knew, but did not have the technical know-how. ☐
4. Knew, had the technical know-how, but did not have the proper guidance to practice it. ☐
5. Knew, had the technical know-how, guidance and was practicing it. ☐

3. Did any of your family members know about Agro-forestry before the beginning of this project?*

* *(Write the total number against the appropriate response)*

1. Did not know anything about Agro-forestry? ☐
2. Knew a bit, had some idea? ☐
3. Knew, but did not have the technical know-how? ☐
4. Knew, had the technical know-how, but did not have the proper guidance to practice it? ☐
5. Knew, had the technical know-how, guidance and was practicing it? ☐

4. Where from did you get to know about Agro-forestry first?*

* *(Tick in the box against the appropriate response)*

1. From some other practicing farmers. ☐
2. From radio/television programmes. ☐
3. From Krishi Vigyan Kendra as part of government sponsored programme. ☐
4. From some other project by private agency before this project. ☐
5. Others (please specify). ☐

Perceptions and Feedback of Non-adopters

1 Given a chance, would you want to join the agro-forestry project	5. Yes, very large (> 80%)	4. Yes, large (60-80%)	3. Yes, moderate (60-40%)	2. Yes, small (40-20%)	1. Yes, very small (1-20%)	0. No (0%)
2. Will you promote agro-forestry systems?	5. Yes, very large (> 80%)	4. Yes, large (60-80%)	3. Yes, moderate (60-40%)	2. Yes, small (40-20%)	1. Yes, very small (1-20%)	0. No (0%)
3. Will you participate in the extension activities of agro-forestry?	5. Yes, very large (> 80%)	4. Yes, large (60-80%)	3. Yes, moderate (60-40%)	2. Yes, small (40-20%)	1. Yes, very small (1-20%)	0. No (0%)
4. In your opinion, the Interactive Bottom up approach adopted by the project was successful?	5. Very high (> 80%)	4. High (60-80%)	3. Moderate (60-40%)	2. Small (40-20%)	1. Very small (1-20%)	0. Not (0%)
5 How do you perceive the overall effectiveness of the project?	5. Very high (> 80%)	4. High (60-80%)	3. Moderate (60-40%)	2. Low (40-20%)	1. Very Low (1-20%)	0. Not at all (0%)

Economic Analysis

Main Crop

Cost of Inputs	
1. Primary Tillage	
2. Cost of Planting Material	
3. Cost of Farm Yard Manure/Fertilizers	
4. Cost of Pesticides	
5. Weeding Costs (twice)	
6. Community Watering Costs	
7. Electricity Charges	
8. Total Labour Costs for all Operations	
Total Input Costs for the Year	

Output	
1. Total yield/acre (in quintals or bags or kgs)	
2. Selling price (in quintals or bags or kgs)	
Total Income	

CHECKLIST FOR IMPLEMENTING NGO

Impact of Biotechnology at Micro Level
Network Project on Agro-forestry, APNLBP

Respondent Number: [ ] **Hamlet:** [ ]

Village [ ] [ ] [ ]

Mandal [ ] [ ] [ ]

District Mehboobnagar | Nalgonda

Implementing NGO [ ]

Name of Enumerator [ ]

Date of Survey [ ]

Data Editing on [ ] **Data Feeding on** [ ]

Institute of Public Enterprise
OU Campus, Hyderabad, Andhra Pradesh

1. **Name of the NGO:**
2. **Brief Profile of the NGO:** (*You can give the related documents, if you please*)
   1. Year of Establishment:

   2. Name of the Director:

   3. Prime Focus Area:

4. Projects Handled:

5. Areas of Operation:

6. Staff Strength:

3. **Years of Experience** of NGO in handling the projects of similar nature:

4. **Number of staff** involved in implementing the Network Project on Agro-forestry:
Highest educational qualification of the staff involved.
Lowest educational qualification of the staff involved.

5. Please share your experiences in implementing the Network Project on Agro-forestry.

6. How was the overall response of the beneficiaries during the implementation of this project?

7. How was the overall response of the non-beneficiaries during the implementation of this project?

Mode of Implementation

1. Did you encourage the formation of a Community-Based Organisation for the implementation of the project?
   1. Yes 2. No
2. If No, Please share the reasons behind it.

3. If Yes,
   (a) Please share the reasons behind it.
   (b) Please draw the organisational diagram of the CBO thus formed.
   (c) Did this CBO, frame its rules and regulations? 1. Yes 2. No
   (d) If Yes, were these rules and regulations formalized and followed by the members?
   (*If possible, please provide the list of rules and regulations*)

Your Opinion

1. The project was able to achieve its stated objectives.
2. The mode of implementation of this project,
   (i) Is excellent, does not need any improvements.
   (ii) Is alright, but needs improvement in certain areas.
3. If it needs improvement, please indicate the areas and corresponding suggestions for it.
4. The project was successful in creating the awareness about Agro-forestry.
5. The farmers have started viewing it as a viable option.

Bibliography

Adato, M. and R. Meinzen-Dick. (2002), Assessing the impact of agricultural research on poverty using the Sustainable Livelihoods Framework. Discussion Paper 89/Discussion Paper 128. Washington, DC: International Food Policy Research Institute.

Altieri, M.A. (1995), *Agro-ecology: The Science of Sustainable Agriculture.* Westview Press, Boulder.

Altieri, M.A., P. Rosset and L.A. Thrupp (1998), *The potential of agro-ecology to combat hunger in the developing world.* 2020 Brief. IFPRI, Washington, DC.

Andersen Pinstrup, Pandya Lorch and Rosegrant (1999), Agricultural Bio-Technology: Meeting the Needs of the Poor, *A Report on the State of Food and Agriculture in 2003-04,* FAO, United Nations, Geneva.

Beets, W.C. (1990), *Raising and Sustaining Productivity of Smallholders Farming Systems in the Tropics.* AgBe Publishing, Holland.

Beyers, L., Y. Ismaël, J. Piesse, and C. Thirtle. (2001), Can Gm-technologies help the poor? The efficiency of Bt cotton adopters in the Makhathini Flats of Kwazulu-Natal. Paper presented at the ISNAR consultation "Biotechnology and Rural Livelihood—Enhancing the Benefits," held in June 2001, The Hague.

Blauert, J. and S. Zadek (1998), *Mediating Sustainability.* Kumarian Press, Connecticut,

Braun, A.R. and E. Van de Fliert (1997), The Farmer Field School Approach to IPM and ICM in Indonesia: User participation. In: Local R&D: Institutionalizing Innovations in Rootcrop Agriculture, 44-64 (UPWARD). UPWARD, Los Baños, Laguna, Philippines.

Chambers, R., A. Pacey and L.A. Thrupp (eds.) (1989), *Farmer First: Farmer Innovation and Agricultural Research.* IT Publications, London.

Chen, S. and M. Ravallion (2005), *How have the world's poorest fared since the early 1980s?,* World Bank, New York

Clark, E.A. (1998), Environmental risks of genetic engineering in field crops. Paper presented to the NAEC workshop "Factoring in the Environment for Decisions on Biotechnology in Agricultural Production," held on 28 July 1998, Ottawa, Canada.

Cohen, J.I. (ed.) (1999), *Managing agricultural biotechnology: Addressing research programme needs and policy implications.* Wallingford, UK: CABI Publishing.

Commission of the European Communities. (2002), "Economic impacts of genetically modified crops on the agri-food sector: A first review." *Working Document Rev.* 2. Directorate General for Agriculture.

Conway, F.R. (1997), *The Doubly Green Revolution.* Penguin Books, London.

Conway, G., and V. Ruttan. (1999), *The Doubly Green Revolution: Food for All in the 21st Century.* Penguin Books, London.

De Steenhuijsen Piters, B. and L. Fresco (1997), Variations on a theme: conceptualizing system diversity with examples from northern Cameroon. In: A. Budelman (ed.) *Agricultural R&D at the Crossroads. Merging Systems Research and Social Actor Approaches,* 33-42. Royal Tropical Institute, Amsterdam. .

Denevan, W.M. (1995), Prehistoric agricultural methods as models for sustainability. *Advanced Plant Pathology* 11:21-43.

Dev Mahendra, S. and Vijay Mahajan (2003), Employment and Unemployment, Ch.15, pp.401-429, in C.H. Hanumantha Rao and S. Mahendra Dev (2003).

Dewalt, B.R. (1994), Using indigenous knowledge to improve agriculture and natural resource management. *Human Organizations* 5: 23-131.

Duffy, M. (2001), Who benefits from biotechnology? Paper presented at the American Seed Trade Association Meeting, held in Chicago, December 5-7, 2001.

Dyson, T. (1999), World food trends and prospects to 2025. In *Proceedings of the National Academy of Sciences* 96: 5929-5936.

Eveleens, K.G., R. Chisholm, E. Van de Fliert, M. Kato, P.T. Nhat, and P. Schmidt (1996), Mid-term Review of Phase III Report. FAO Inter-country Programme for the Development and Application of Integrated Pest Control in Rice in South and South-east Asia, P.O. Box 3700 MCPO, 1277 Makati, Metro Manila, Philippines.

Ewen, S.W.B. and A. Pusztai. (1999), Effects of diets containing genetically modified potatoes containing *Galanthus nivalis* lectin on rat small intestines. In *The Lancet* 354: 9187.

Falck-Zepeda, J.B., G. Traxler, and R.G. Nelson. (2000), Surplus distribution from the introduction of a biotechnology innovation. In *American Journal of Agricultural Economics* 82: 360-369.

Fano, H., O. Ortiz, and T. Walker (1996), Peru: Inter-Institutional Cooperation for IPM. In: L.A. Thrupp (ed.) *New Partnerships for Sustainable Agriculture,* 85-98. World Resources Institute, Washington.

FAO (2003), *Agriculture Biotechnology: Meeting the Needs of the Poor,* The State of Food and Agriculture, Food and Agriculture Organization of the United Nations, Rome.

FAO (2000), World Agriculture: *Towards 2015/30, Technical Interim Report.* Food and Agriculture Organization of the United Nations, Rome.

Francis, C.A. (1986), *Multiple Cropping Systems.* MacMillan, New York.

Gladwin, C. and K. Truman (1989), *Food and Farm: Current Debates and Policies.* University Press of America, Lantham, MD.

Gliessman, S.R. (1998), *Agro-ecology: Ecological Process in Sustainable Agriculture.* Ann Arbor Press, Michigan.

Gulati Ashok and Kathleen Mullen (2003), Responding to Policy Reform: Indian Agriculture in 1990s and After, Working Paper No.189, Stanford University; Stanford.

Haverkort, B., J. Van der Kamp and A. Waters-Bayer (eds.) (1991), *Joining Farmers' Experiments. Experiences in Participatory Technology Development.* IT Publications, London. pp. 269.

Ho, M.W. (1999), *Genetic Engineering—Dream or Nightmare? Turning the Tide on the Brave New World of Bad Science and Big Business.* 2nd ed. Dublin: Gateway, Gill and Macmillan.

Huang, J., R. Hu, C. Pray, F. Qiao, and S. Rozelle. Biotechnology as an alternative to chemical pesticides: A case study of Bt cotton in China. Paper presented at the ISNAR Consultation Biotechnology and Rural Livelihood—Enhancing the Benefits, held in June 2001, The Hague.

Jimenez-Osornio, J. and S. del Amo (1986), An intensive Mexican traditional agro-ecosystem: the chinampa. Proc. 6th International Scientific Conference IFOAM, Santa Cruz, California.

Kaeppler, H. (2000), Food safety assessment of genetically modified crops. In *Agronomy Journal* 92: 793-797.

Kerr, J. and S. Kolavalli. (1999), Impact of agricultural research on poverty alleviation: Conceptual framework with illustrations from the literature. Discussion Paper 56. Washington, DC: International Food Policy Research Institute.

Kuiper, H.A., H.P.J.M. Noteborn, and A.A.C.M. Peijnenburg. (1999), Adequacy of methods for testing the safety of genetically modified foods. In *The Lancet* 354: 1315-1316.

Lampkin, N. (1992), *Organic Farming.* Farming Press, Ipswhich, England.

Lappe, F.M., J. Collins and P. Rosset. (1998), *World Hunger: Twelve Myths.* Grove Press, New York.

Lipton, M. and R. Longhurst, (1989), *New Seeds and Poor People.* John Hopkins University Press, Baltimore.

Losey, J.E., L.S. Rayor, and M.E. Carter. (1999), Transgenic pollen harms Monarch larvae. In *Nature* 399: 214.

Marra, M.C., P.G. Pardey and J.M. Alston. "The payoffs to agricultural biotechnology: An Assessment of the Evidence." Discussion Paper 87. International Food Policy Research Institute, Washington, DC.

Meinzen-Dick. R.S. (2001), Measuring the livelihood impact of agricultural research. Paper presented at the ISNAR consultation "Biotechnology and Rural Livelihood—Enhancing the Benefits, held in June 2001, The Hague.

Moore-Lappé, F.J. Collins, P. Rosset, and L. Esparza. (1998), 12 *Myths about Hunger based on World Hunger: 12 Myths*. 2nd ed. New York: Grove/Atlantic and Food First Books.

National Academy (2000), *Transgenic Plants and World Agriculture.* Washington, DC: National Academy Press.

Netting, R. McC. (1993), *Smallholders, Householders.* Stanford University Press, Santford, CA.

Nordlee, J.A., S.L. Taylor, J.A. Townsend, L.A. Thomas, and R.K. Bush. (1996), "Identification of a Brazilnut Allergen in Transgenic Soybeans." *New England Journal of Medicine* 334: 688-692.

Nordström, H. (2000), "Trade, Income Disparity and Poverty: An Overview". In: Ben-David, D., Nordström, H. and Winters, L.A., *Trade, Income Disparity and Poverty.* Geneva: World Trade Organization, Special Studies No. 5.

Ortega, E. (1986), *Peasant Agriculture in Latin America.* Joint ECLAC/FAO Agriculture Division, Santiago.

Paarlberg, R.L. (2000), Governing the GM Crop revolution. IFPRI Discussion Paper 33. Washington, DC: International Food Policy Research Institute.

Pearse, A. (1980), *Seeds of Plenty Seeds of Want; Social and Economic Implications of the Green Revolution.* Oxford University Press, New York.56.

Pingali, P.L., M. Hossain and R.V. Gerpacio. (1997), *Asian Rice Bowls: The Returning Crisis.* CAB International, Wallinford, UK.

Pinstrup-Andersen, P., R. Pandya-Lorch, and M.W. Rosegrant. (1999), World food prospects: Critical issues for the early Twenty-First Century. *2020 Vision Food Policy Report.* Washington, DC: International Food Policy Research Institute.

Posner, J.L. and M.F. McPherson (1982), *Agriculture on the Steep Slopes of Tropical America.* World Development 10:341-53.

Pretty, J. (1995), *Regenerating Agriculture.* World Resources Institute. Washington, DC.

Radhakrishna, R. (2002), "Agricultural growth, employment and poverty: A policy perspective", *The Indian Journal of Labour Economics,* Vol. 45, No.1, 3-16.

Ramana Murthy, R.V. and Wendy K. Olsen (2000), 'Contract labour and bondage in Andhra Pradesh (India)', *Journal of Social and Political Thought,* Vol. 1, No. 2, June.

Reddy, Narasimha. D (2002), "Changing agrarian relations and rural labour: Certain emerging issues", *The Indian Journal of Labour Economics,* Vol. 45, No.1, 47-68.

Reddy, Ratna V. (2003), Irrigation: Development and Reforms, Ch.6 in C.H. Hanumantha Rao and S. Mahendra Dev (2003).

Reddy, Sudhakar, S. S. Galab and P. Padmanabha Rao (2003), Trends and Determinants of Poverty, Ch.16, pp.430-461 in C.H. Hanumantha Rao and S. Mahendra Dev (2003).

Reinjtes, C., B. Haverkort and Ann Waters-Bayer (1992), *Farming for the Future.* MacMillan, London.

Richards, P. (1985), *Indigenous Agricultural Revolution.* Westview Press, Boulder.

Rogers, E.M. (1995), *Diffusion of Innovations.* 2nd ed. Free Press, New York.

Röling, N. (1988), *Extension Science: Information Systems in Agricultural Development.* Cambridge University Press, Cambridge.

Röling, N. and E. Van de Fliert (1998), Introducing Integrated Pest Management in Rice in Indonesia: A Pioneering Attempt to Facilitate Large Scale Change. In: N.G. Röling and M.A.E. Wagemakers (eds.) *Sustainable Agriculture: Participatory Learning and Action,* 153-171. Cambridge University Press, Cambridge.

Sanders, W.T. (1957), Tierray Agua: A study of the ecological factors in the development of Meso-American Civilizations. Ph.D. Dissertation, Harvard University.

Scoones, I. and J. Thompson (1994), *Beyond Farmer First Rural People's Knowledge, Agricultural Research and Extension Practice.* IT Publications, London. p. 310.

Sekhar, C.S.C. (2003), Volatility of Prices—An analysis of major international and domestic markets, Working Paper No.103, Indian Council for Research on International Economic Relations; New Delhi.

Shelton, A.M., J.-Z. Zhao, and R.T. Roush. (2002), Economic, ecological, food safety, and social consequences of the deployment of Bt transgenic plants. In *Annual Review of Entomology* 47: 845-81.

Shiva, V. (1991), *The Violence of the Green Revolution: Third World Agriculture, Ecology and Politics.* Third World Network, Pengany, Malaysia 56.

Shiva, V. (2000), *Tomorrow's Biodiversity (Prospects for Tomorrow).* London: Thames and Hudson.

Singh, Lakhawinder (2006), Innovations, High-Tech Trade and Industrial Development Theory, Evidence and Policy, UNU-WIDER Anniversary Conference, 'WIDER Thinking Ahead: The Future of Development Economics', Helsinki

Subrahmanyam, S. and P. Satya Sekhar (2003), Agricultural Growth: Patterns and Prospects, Ch.8, pp.221-245, in C.H. Hanumantha Rao and S. Mahendra Dev (2003).

Subrahmanyam, S. (2003), Regional Disparities: Causes and Remedies, Ch.19, pp.517-546, C.H. Hanumantha Rao and S. Mahendra Dev (2003).

Thrupp, L.A. (1996), *New Partnerships for Sustainable Agriculture.* World Resources Institute, Washington, DC.

Toledo, V.M., J. Carabias, C. Mapes, and C. Toledo. (1985) *Ecologiay Autosuficiencia Alimentaria.* Siglo Veintiuno Editores, Mexico City.

Traxler, G., S. Godoy-Avila, J.B. Falck-Zepeda, and J. de J. Espinoza-Arellano. The impact of Bt cotton in Mexico. Paper presented at the ISNAR Consultation Biotechnology and Rural Livelihood—Enhancing the Benefits, held in June 2001, The Hague.

UNDP (1999), *Human Development Report 1999,* UNDP/Oxford University Press, New York.

Van de Fliert, E. (1998), Integrated Pest Management: Springboard to Sustainable Agriculture. In: G.S. Dhaliwal and E.A. Heinrichs (eds.) *Critical Issues in Insect Pest Management,* 250-266. Commonwealth Publishers, New Delhi (India).

Van de Fliert, E. and A.R. Braun (1999), From problem to impact: A conceptual model for integrative, farmer participatory research for sustainable agriculture. Submitted to Experimental Agriculture.

Van Veldhuizen, L., A. Waters-Bayer and H. de Zeeuw (1997), *Developing Technology with Farmers. A Trainer's Guide for Participatory Learning.* Zed Books Ltd., London. pp. 230.

Wilken, G.C. (1987), *Good Farmers: Traditional Agricultural Resource Management in Mexico and Guatemala.* University of California Press, Berkeley.

World Bank (2000), *World Development Report 2000/2001: Attacking Poverty.* World Bank, Washington.

World Bank (2004), *World Development Report 2005,* Oxford University Press: New York.

Index